METEOROLOGY FOR WIND ENERGY

METEOROLOGY FOR WIND ENERGY

AN INTRODUCTION

Lars Landberg

DNV GL, Copenhagen, Denmark

WILEY

This edition first published 2016

© 2016 John Wiley & Sons Ltd

Registered office

John Wiley & Sons Ltd, The Atrium, Southern Gate, Chichester, West Sussex, PO19 8SQ, United Kingdom

For details of our global editorial offices, for customer services and for information about how to apply for permission to reuse the copyright material in this book please see our website at www.wiley.com.

Library of Congress Cataloging-in-Publication Data applied for.

A catalogue record for this book is available from the British Library.

ISBN: 9781118913444

Set in 10/12.5pt Times by Aptara Inc., New Delhi, India

1   2016

To my family: My wife, Frances, and our two big boys, Marcus and Lucas, thank you so much for your support and understanding.

Contents

About the Author

Lars Landberg (born 1964) has been working in the wind energy field since 1989: the first 18 years at Risø National Laboratory (now DTU Wind), a research lab in Denmark, and since then at Garrad Hassan (now DNV GL), a global wind energy consulting company. His main areas of expertise are wind resource estimation and short-term prediction of wind power. Lars has taught meteorology-related courses to the wind energy industry since the first Risø WAsP course in 1991. Lars has a PhD in physics and geophysics from the University of Copenhagen and an MBA from Warwick Business School, United Kingdom.

Foreword

Lars Landberg is right that there are too many of us using the wind without understanding it. Lars' book will help address that problem and it will entertain as well as educate. He has already taught much of the wind industry about meteorology and, through this book his undoubted didactic talents will reach a much wider audience. Part of his job at Garrad Hassan was to encourage free thinking and who better to do that than someone who cut his meteorological teeth on the Martian weather? You can be sure that he will bring a refreshing approach to this difficult subject.

Our generation has witnessed an extraordinary transition in the electricity industry. Wind used to occupy a position on its fringes where we were patronised as eccentric and naïve ideologists – now this fantastic industry is part of the main stream driven by clean and powerful wind. Being able to make a reasonable job of predicting both the behaviour of the wind and the resulting behaviour of the turbines and wind farms has been an essential part of that success.

I have known Lars Landberg for a long time, first as a fellow wind energy enthusiast, then as a colleague and a friend. He was a member of the Garrad Hassan board and it was certainly stimulating to have a board member who had written a book called 'Strategy: No Thanks!'. From that title you can deduce that this book will have unexpected dimensions to it. I came across the Ekman spiral (my first acquaintance with meteorology) while writing my undergraduate thesis in 1973. I am ashamed to say that I did not know that Ekman was a Swede until I opened this book.

The wind is the free, clean fuel that distinguishes our industry from the expensive dirty stuff that others use but it is also the source of the loads that break the turbines. My meeting with Ekman was in the same year that I made my first wind mill alas destroyed by the wind very quickly. The wind therefore commands both our gratitude and our respect. This book will help us all to deliver that respect by initiating a wider understanding of the wind. Well done Lars!

Andrew Garrad
23 March 2015

Preface

Having taught various aspects of meteorology to all kinds of people working in wind energy since 1991, I was very happy to accept the kind offer from Wiley to write a book on the subject. Wind energy, probably like many other 'new' fields, not having a dedicated and established way of teaching the subject (like e.g. medicine), sees people entering the field from a multitude of technical and academic backgrounds, but rarely from meteorology. Often, therefore, the meteorology that people know, is taught to them on a need-to-know basis when they join their companies. This means that I, over the years, have met many people who have lacked the fundamentals of meteorology (general as well as the more specific area of boundary-layer meteorology), and when attending the various courses that I have taught, have expressed great satisfaction in knowing the basics, even though this is 'only' nice to know.

When writing this book, I have had the above-mentioned varied technical and academic background of the reader in mind, so do not expect a PhD-level book, but expect to be able to understand the meteorological basics of what you do every day. As a general principle, I have gone for hand-waving understanding, rather than strict physical explanations, so some corners have inevitably been cut, but I have done my utmost to make sure you are being made aware of this when it happens. In doing so, I might also have made some errors (I hope not of course), and the website (larslandberg.dk/windbook, see QR code in margin) will be updated as soon as I have been made aware of any.

Another important point to stress is that the atmosphere is a **complex, interconnected, three-dimensional physical system**. This means that in order to understand it fully (which nobody does, actually), one needs to solve the entire set of physical laws that govern this system, this is first of all difficult, but secondly it only explains the atmosphere to the level of a black box. So what we will do in this book is that we will zoom in on different aspects, like the low-pressure system, which of course only is a reflection of the laws of motion, temperature and humidity, but dissecting it into fronts helps us understand it much better, in a very useful hand-waving way.

The general idea is that I use simple maths to explain things, so if I can ask you not to be afraid of this, I promise you that I shall guide you through the various lines of reasoning in a safe manner! Also, it is a good idea to fire your favourite spreadsheet up; it is good for looking at numbers and plotting them. In the more difficult or important parts of the text, I have inserted exercises directly in the text, and in many cases I have also done those exercises as part of the text. Despite the fact that you have the answer right after the exercise, please try to have a go at it first; your understanding of the subject will benefit immensely. If you absolutely hate maths, there is also a path through the book, where you can just read the text and not do any of the exercises, it will not be as fun, of course, but it can be done.

For people interested in digging just one layer deeper, I have inserted boxes in various places along the way. These boxes shall focus on one topic, often quite technical, and explain a bit more about it. If the subject does not interest you, you should be able to skip these boxes without disturbing your overall understanding of the text.

To make the various equations a bit more personal, I have also, in many places, inserted a brief description of the person or persons behind the equation. There are a lot of men in those boxes, despite the fact that I have tried to get both sexes represented.

The structure of the book is that we start with the meteorological basics, where the scene will be set and some fundamentals of general meteorology will be covered. We will then discuss measurements, where some measurement philosophy, theory and basics will be covered. Whether measurements are carried out by means of a mast or a remote-sensing device, the result is a picture of the vertical structure of the atmosphere at the measuring location, we will therefore cover the theory of this vertical structure, often called the *wind profile*. The wind profile is the result of atmospheric flow on many scales and understanding the flows makes us able to infer the profiles (and in many cases vice versa). Two further aspects will then be covered: turbulence and wakes.

Having gone from measurements via flow on all scales to wakes, the circle is complete in some sense; however, in order to understand these topics in more detail, we have introduced a great number of models and it is appropriate to have a chapter on the general aspects of *modelling* (Chapter 8), where no new models will be introduced, but the more philosophical and theoretical aspects of modelling will be discussed.

Lars Landberg
Copenhagen, Denmark

Kaze: The Japanese kanji for 'wind' by Shigemi Nakata

Acknowledgements

The material for this book has been built up in my mind ever since I started working in wind energy, which is 26 years ago. I have met so many people who have taught, inspired and discussed with me, and it is, unfortunately, impossible to list them all here. But I would like to highlight and deeply thank the following people:

At the University of Copenhagen, Aksel Walløe Hansen has been the supervisor on my Masters as well as my PhD thesis: you have always been a great supervisor, and always asked difficult questions that have brought me forward!

At Risø (now DTU Wind), I would like to thank the entire Meteorology Group, but in particular, the members of the WAsP team: Niels Gylling Mortensen, Ole Rathmann, Lisbeth Myllerup and Rikke Nielsen. We have worked together on many of the ideas that you find in this book; thank you for a great time together. Outside the WAsP team my good colleagues Søren E. Larsen, Erik Lundtang Petersen, Leif Kristensen, Jakob Mann and Hans E. Jørgensen also deserve a great thank you, again many of the ideas in my head (and in this book) originated from discussions with you guys.

At Garrad Hassan (now DNV GL), I would in particular like to thank Andrew Garrad. He was the one who saw a business in teaching meteorology to wind energy people and prompted me to develop the course that many of the ideas in this book are based on.

Before starting the book, I would also like to thank the five anonymous reviewers for taking the time to look through the original proposal for the book and making some very useful suggestions, many of which have been included. Wolfgang Schlez, for help in understanding the depths of the Ainslie wake model and also for reading through the wakes chapter and providing very valuable input. Jean-François Corbett, for meticulously reading through the Local Flow and Modelling chapters and also producing very valuable feedback.

I would also like to thank Kurt S. Hansen and winddata.com, for allowing me to use the data for the exercises and examples in the chapter on Measurements.

Thank you also to Søren William Lund, who so readily and without warning lined up a lot of instruments for a 'photo session'.

Shigemi Nakata, for the beautiful kanji that you find on the front cover and on page xvi: domo arigato gozaimasu!

Andrew Garrad has also very kindly written the foreword to this book: thank you very very much, Andrew.

And a general thank you to all the copyright holders who have all – without exception – replied swiftly and positively to my many requests for use of their material.

The author thanks the International Electrotechnical Commission (IEC) for permission to reproduce information from its International Standard IEC 61400-1 ed.3.0 (2005). All such extracts are copyright of IEC, Geneva, Switzerland. All rights reserved. Further information on the IEC is available from www.iec.ch. IEC has no responsibility for the placement and context in which the extracts and contents are reproduced by the author, nor is IEC in any way responsible for the other content or accuracy therein.

List of Abbreviations

ABL	Atmospheric boundary layer
agl	above ground level
CFD	Computational fluid dynamics
DNS	Direct numerical simulation
ENSO	El Ninõ Southern Oscillation
GPS	Global Positioning System
GTS	Global Telecommunication System
IBL	Internal boundary layer
IEC	International Electrotechnical Commission
IPK	International Prototype of the Kilogram
ISA	International Standard Atmosphere
ISO	International Organization for Standardization
ITCZ	Inter Tropical Convergence Zone
LES	Large eddy simulation
MCP	Measure–Correlate–Predict
NAO	North Atlantic Oscillation
NASA	National Aeronautics and Space Administration
NCAR	National Center for Atmospheric Research
NCEP	National Centers for Environmental Prediction
NOAA	National Oceanic and Atmospheric Administration
NWP	Numerical Weather Prediction
PBL	Planetary boundary layer
RANS	Reynolds-averaged Navier–Stokes
SAR	Synthetic Aperture Radar
SOI	Southern Oscillation Index
SST	Sea surface temperature
UN	United Nations
WAsP	Wind Atlas Analysis and Application Programme
WMO	World Meteorological Organisation
WRF	Weather Research and Forecasting

1

Introduction

This book is about the meteorological aspects of wind energy.[1] There are mainly two areas which are 'affected by the wind': wind resource estimation and loads. This book will focus on the meteorological aspects of wind resources, but loads will also be discussed, mainly in the chapter on turbulence (Chapter 6).

The structure of the book is, as a general principle, that each chapter will describe one subject and start with the basics, then the intermediate and lastly the more advanced topics of each subject. Depending on what you are interested in, in the various subject areas, you can stop when you have reached the level you are looking for. At the end of each chapter I have put some exercises: some are quite basic, just using what you have learnt, some more technical requiring calculations, etc. and some are more open-ended, just to get you to think about things. In other cases, where it fitted better with the flow of the chapter, I have put exercises right in the text; they are intended for you to pause and think. Some of the exercises are solved in the text following the exercise, because the point is so important, that you need to know the answer in order to proceed, others are solved in the Answers appendix. The answer to each of the exercises can be found in the back (Appendix B). To solve the exercises all you need is pen, paper and a pocket calculator, but you will make your life much easier if you use a spreadsheet.

In many places in the text you will find *references* to books, articles, websites, etc. These are meant, firstly to follow the academic tradition of acknowledging other peoples' work, but also to point you in a specific direction, where you will be able to learn more. In this day and age, many things can be found for free on the internet. This is, unfortunately, not always the case with references, so consulting a given reference might not always be as easy as just a click away.

As a general principle, I have tried to make the book a journey, and one where I walk along with you. This might have led to a bit of a chatty style in places, but I think overall, by presenting the subjects in this way, your learning should be helped significantly.

---

[1] But I am sure most of the chapters can also be used as a good general introduction to the field of general meteorology and in particular, the field called boundary-layer meteorology.

Meteorology for Wind Energy: An Introduction, First Edition. Lars Landberg.
© 2016 John Wiley & Sons, Ltd. Published 2016 by John Wiley & Sons, Ltd.

The sequence of the chapters is as follows: First we start with a chapter on *meteorological basics* (Chapter 2) where the scene will be set and some fundamentals of general meteorology will be covered. We will then discuss *measurements* (Chapter 3), where some measurement philosophy, theory and basics will be covered. Whether measurements are carried out by means of a mast or a remote-sensing device, the result could be a picture of the vertical structure of the atmosphere at the measuring location, the following chapter (Chapter 4) therefore covers the theory of this vertical structure, often called the *wind profile*. The wind profile is the result of atmospheric *flow* on many scales, and in many ways this is like being shown the needle and then asked to describe the haystack! But understanding the various flows makes us able to infer the profiles (and in many cases vice versa), so these flows will be described in the next chapter (Chapter 5). Two further aspects will then be covered: *turbulence* (Chapter 6), which is quite a technical subject, but I will do my best to explain and then, finally, *wakes* (Chapter 7), that is the downstream reduction of the wind speed due to the wind turbine.

Having gone from measurements via flow on all scales to wakes, the circle is complete in some sense; however, in order to understand all of this in more detail, various models have been introduced and it is therefore appropriate to have a chapter on the general aspects of *modelling* (Chapter 8), where no new models will be introduced, but the more philosophical and theoretical aspects of modelling will be discussed.

Finally, some conclusions will be drawn (Chapter 9) and the list of references will be given. As a quick way of summarising the most important formulae and points, you will, in Appendix A, find a 'cheat sheet', which can be used as a quick reference guide. In Appendix B answers to all the exercises will be given.

The book also has an accompanying website (larslandberg.dk/windbook, see QR code in margin), where data for some of the exercises can be found. Here you will also find a few videos, an area where I will discuss each chapter further (in cases where new information becomes available), list the errata (if any!) and finally I will post news items of various kinds there as well.

Once you have read this book, it is my goal and hope that you will have a basic understanding of all the relevant areas of meteorology needed for working in wind energy. As mentioned I will have cut some corners, so I also hope that you will be in a position to continue your learning in the areas that particularly interest you.

You will also realise, as you read through the chapters, that most questions we can ask to try to understand any of the topics discussed in this book can rarely be answered with a straight 'yes' or 'no'; instead the answer is often: 'It depends', and one of my main goals of the book is also to enable you to qualify that very simple statement, to ask the right questions and to be able to state what this or that actually depends on.

You might as well get used to the fact that I will ask you to work with me solving different exercises as we go along, so here is the first one, where we, step by step, will derive one of the fundamental relations in wind energy:

Exercise 1.1 *Imagine for a short while that you have the front frame of a football goal dangling in mid air (might be a bit difficult to, but the idea is that we want to look at a box of air), and furthermore, that it is exactly 1 × 1 m. Imagine also that the wind is blowing through the frame at a speed of 10 m/s. How many cubic metres of air will pass through the goal in 1 s?*

As mentioned above, the expression we will get out at the end of these derivations is quite fundamental, so I will solve the exercise here; however, please give it a try yourself first. As you will see, I have broken this first exercise down into a lot of small steps, to ensure that you can follow me, and get used to this way of working.

The area is 1×1 m, and at a wind speed of 10 m/s, a box 10 m long will pass through in 1 s, viz

$$10 \times 1 \times 1 = 10 \text{ m}^3 \tag{1.1}$$

of air will pass through.

Exercise 1.2   *Assume the density is ρ kg/m³, how many kg of air does then blow through the goal?*

We have 10 m³ of air, so we get the mass of the air going through the goal, m, to be:

$$m = 10 \cdot \rho \tag{1.2}$$

Note that we are not using the actual value of ρ (which is around 1.225 kg/m³), since we are trying to get to a general expression.

Exercise 1.3   *Assume now that the wind speed is u m/s (instead of the 10 m/s), what would the mass, m, of the air going through the 1 m² goal frame be?*

Generalising what we have just found, we get:

$$m = u\rho \tag{1.3}$$

We now have the first part of an expression that we are working on deriving, the second part has to do with the kinetic energy, so

Exercise 1.4   *What is the definition of kinetic energy?*

This can be found in many books on fundamental physics and it is:

$$E_{kin} = \frac{1}{2}mu^2 \tag{1.4}$$

We are now able to answer the question that I have been looking for an answer to:

Exercise 1.5   *Combine what we have found out about the mass of the air going through the goal frame with the equation for kinetic energy, to see how much kinetic energy goes through 1 m² of air.*

Combining the two relations, we get

$$E_{kin} = \frac{1}{2}mu^2 = \frac{1}{2}(\rho u)u^2 = \frac{1}{2}\rho u^3 \qquad (1.5)$$

Which is the amount of kinetic energy that blows through 1 m^2.

This is one of the more fundamental equations in wind energy, showing that the energy is propositional to the *cube* of the wind speed. A different way of explaining what this equation says, is that looking at 1 m^2 of the rotor plane of a wind turbine, this is the maximal energy available for the wind turbine to convert into electricity.[2]

You are about to go on a tour through quite a few different topics, both with regard to breath of coverage, but also mathematically, some topics are easy to approach, but others, like turbulence, are much more difficult. But, if you just allow yourself to at least read through those parts, too, you will gain a bit more understanding, and when the topic pops up at a later stage, you will find the terms familiar and your understanding will gradually increase.

---

[2] There is a law, called Betz's law, that says that no wind turbine can capture more than 16/27 ($\approx$59.3%) of the kinetic energy in the wind.

2

Meteorological Basics

In this chapter, we will learn a bit about the atmosphere, starting with finding out why the wind blows! Then a few things about the weather and such basic things. Most of what I will describe in this chapter can be found in many a lecture note and book on basic meteorology; however, by collecting and including it here, you will have everything in the one place. I have put references to many of these books in the following. My personal favourite is the – now quite old – book written by Robin McIlveen (McIlveen, 1986), but there are many more books available and I have listed a few more on the website (larslandberg.dk/windbook).

2.1 Why Does the Wind Blow?

Exercise 2.1 *Sorry to start this chapter with a question! But could I ask you to pause for just a minute and think about why the wind blows? (There are two types of exercises in this book: the first (like the present one) where the answer comes right after the exercise and the second where you can find the answer in Appendix B. In the first case, you of course need to be very disciplined and make sure you pause and think before you read on! I will try to remind you as well.)*

The most fundamental answer to this very basic question is *not* that it is due to pressure differences (it is of course that, too), and *not* that it is because of local temperature differences (it is of course, too). I mention these two explanations since they, in my own experience, are what most people suggest as the first answer. The answer, at the most fundamental level, is that the Poles receive less energy per unit area than the Equator. If you take a look at the very simplified picture of the Earth in Figure 2.1, you can see that the amount of energy that each square metre receives at the Poles is less than on the Equator because the area is stretched out at the Poles. This leads to the heat 'building up' at the Equator. Such a build up is not 'allowed' in physical systems, and the equatorial warmer air will rise, and the colder polar air will sink,

Meteorology for Wind Energy: An Introduction, First Edition. Lars Landberg.
© 2016 John Wiley & Sons, Ltd. Published 2016 by John Wiley & Sons, Ltd.

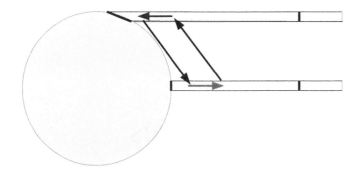

Figure 2.1 The Sun's rays hitting the Earth, more energy per area at the Equator, less at the Poles. Hot air rises, cold air sinks. Non-rotating case. Imagine the Sun to be very far to the right

developing a flow pattern with the one and only function of transporting this excess heat from the Equator to the Poles (i.e. to strive to get the system into equilibrium).[1]

Looking at weather maps you might realise that such an Equator–Pole flow is rarely (if ever) observed in reality; this is due to one very important fact that we so far have not taken into account, the *rotation* of the Earth. This rotation introduces the Coriolis force (see Section 2.3.2), which makes the wind veer (turn), creating a significantly more complicated picture. A simplified version of this can be found in Figure 2.2. It is important to stress that the most important part of the global weather circulation is not the various cells in the illustration (despite their prominence there), but the westerlies and the trade winds.

A snapshot of the weather can be seen in Figure 2.3, as can be seen, reality is quite a bit more complicated!

To summarise: the wind blows on the Earth due to the different amounts of radiation received at the Equator and the Poles. The overall wind patterns we see are due to this and the fact that the Earth is rotating.

We have now answered the most fundamental question, why does the wind blow and will continue with some other basic facts about the atmosphere, starting with the vertical structure.

2.2 The Vertical Structure of the Atmosphere

The vertical structure of the atmosphere is given in Table 2.1. We live in the troposphere and most, if not all, of the weather happens here, too. When we fly, we sometimes enter the stratosphere. Every time a '-sphere' ends, there is a '-pause', for example, between the troposphere and the stratosphere you will find the tropopause. In the beautiful photograph in Figure 2.5 one can just make out the troposphere (the first grey area from the left), the whiter bit is the stratosphere and the other grey area is the mesosphere. As we get further and further away from the Earth's surface (i.e. into the meso-, thermo- and exosphere), the layers get more

---

[1] This same process is also what happens in the oceans.

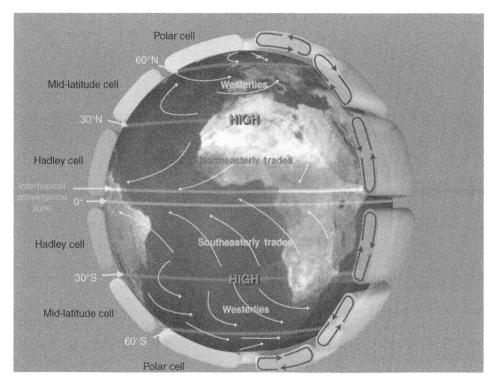

Figure 2.2 A simplified picture of the Earth's global atmospheric circulation. Note the Inter Tropical Convergence Zone (ITCZ), we will need it later. Source: NASA http://sealevel.jpl.nasa.gov/files/ostm/6_celled_model.jpg

and more esoteric of course, and as indicated by the very low pressures there, very little can actually be found there.

Once we are finished with this chapter, we will forget about most of the layers and only focus on the part of the atmosphere closest to the surface of the Earth, that is, we are now well within the troposphere. This lower part is known as the planetary boundary layer (PBL), also known as the atmospheric boundary layer (ABL). This layer goes up to about 1 km, and can again be subdivided into the following layers:

Surface Layer: This is where the logarithmic wind profile is valid (see Chapter 4). The height of this layer varies primarily with the time of day and atmospheric stability (see Section 4.8), and it can be as low as less than 100 m up to, what is a more typical value, a couple of 100 m. Looking at this from a wind turbine perspective, you can see that the rotor sometimes will be entirely inside the surface layer, and other times either partly or even entirely in the Ekman layer. The surface layer is also sometimes called the Prandtl layer or the constant flux layer.

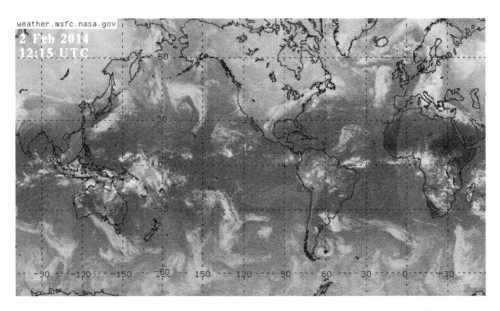

Figure 2.3 A snapshot of the global weather as seen from the Global Geostationary Weather Satellites (infrared channel, i.e. white areas are clouds). Source: NASA http://weather.msfc.nasa.gov/GOES/globalir.html. Data courtesy of the NCEP Aviation Weather Center located in Kansas City, Missouri, United States

Table 2.1 The different layers of the atmosphere. Between each sphere a 'pause' can be found, for example, between the troposphere and the stratosphere, we find the tropopause. Temperature ranges are approximate. From the thermosphere and up, temperature takes on a different meaning, since the pressure and densities are so low, that the definition of temperature falls back to the basic one of the velocity of the molecules. See Figure 2.4 for more on the variation of temperature with height

Name	Start height (km)	Pressure (hPa)	Temperature (°C)	Comments
Troposphere	0	1013	Falling, 20 to −50	This is where we live
Tropopause				Jet stream
Stratosphere	11	226	Rising, −50 to 0	Temperature increases due to the presence of ozone
Stratopause				
Mesosphere	47	1	Falling, 0 to −90	
Mesopause				
Thermosphere	85	≈0	Rising, −90 and up	The space shuttle flew here
Thermopause				
Exosphere	700	≈0		
Exopause				

Height and pressure values from the ISA (International Standard Atmosphere, ISO 2533:1975), up to the mesosphere.

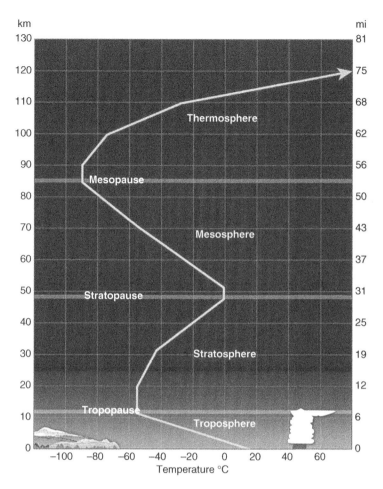

Figure 2.4 The vertical variation of temperature in the atmosphere. The temperatures are typical values. Approximate heights are given in either km (left *y*-axis) or miles (right *y*-axis). For a connection to the pressure levels, please refer to Table 2.1. Source: NOAA. Reproduced with permission of NOAA

> **Ekman layer:** This layer is above the surface layer and is characterised by the fact that there is a three-component force balance between the pressure gradient, the friction and the Coriolis forces (much more about this in Section 2.3.3).

On top of the PBL, we find the 'free atmosphere', and one definition of it is that, this is the layer where the frictional force (see Section 2.3.2) becomes negligible, in other words, the layer is free from the influence of the surface. The different layers of the lower atmosphere are shown in Figure 2.6.

Recently (Banta et al., 2013), people have also been talking about the so-called *rotor layer*, that is, the vertical layer covered by the wind turbine rotor disk. This is of course a nonphysical and not very clear definition, since the sizes of the wind turbines vary a lot according to the

Figure 2.5 The Earth's atmosphere and a space shuttle. One can make out the troposphere (first grey area from the left), the stratosphere (the white area) and the mesosphere (the second grey area). Source: NASA, Photo ID: ISS022-E-062672

turbine type. However, I do think that it makes some sense, since as you can see in Figure 2.6, a standard turbine (height around 100 m, rotor disc about the same) will in many cases be both in the surface layer and in the beginning of the Ekman layer.

Before we dive into the boundary layer, it would seem appropriate to answer the question: when does the atmosphere of the Earth *end*? There are different answers to this questions, but the most common one is that it is about 10,000 km (NASA, 2014). The atmosphere does not end sharply as such, of course, there are just very very few molecules at this height, so it just gradually changes into (empty) outer space.

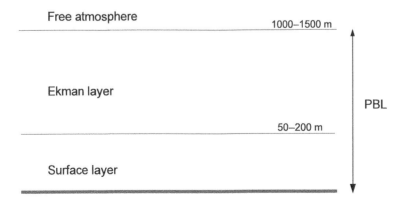

Figure 2.6 The different layers of the lower part of the troposphere, called the planetary boundary layer (PBL). Nearest to the surface we find the surface layer, then the Ekman layer and on top of that the free atmosphere. Typical heights are given. The height of the PBL is sometimes denoted by z_i

2.3   Atmospheric Variables and Forces

In this section, I will briefly take you back to Physics 101, sorry! The basic atmospheric variables will be introduced and likewise the fundamental forces that are of significance to flow in the atmosphere. The reason for doing this is that we, going forward, time and time again, will need these when describing various aspects of the flow and the atmosphere.

2.3.1   Atmospheric Variables

Basically there are the following variables that we need to take into account when discussing the flow and other related aspects of the atmosphere.

> **Pressure,** usually denoted by p, unit Pa ($=N/m^2$) (sometimes the non-SI unit bar, b, and especially the millibar, mb ($=hPa$), is still used)
>
> **Temperature,** usually denoted by T, unit K or °C
>
> **Density,** usually denoted by ρ, kg/m^3
>
> **Velocity,** usually denoted by u, v, w for the x, y, z-component, respectively. The velocity vector is denoted by $\vec{V}$, unit m/s
>
> **Humidity,** usually denoted by q, unit g (water vapour)/kg (dry air).

As you can see, we are only concerned with a handful of variables, and once we zoom in on the wind farm scale, the set gets even smaller. Note that there are many different kinds of temperature, and when we talk about atmospheric stability (in Section 5.6), we will need to be more precise in its definition.

2.3.2   Atmospheric Forces

We have now established the set of variables that are important in meteorology and we will continue with the forces where the following list governs the flow in the atmosphere:

- pressure gradient force
- frictional force
- Coriolis force
- gravitational force.

Strictly speaking, the Coriolis force is not a real force (it is a so-called fictitious force), and there is also something funny about the gravitational force (see later in this section), but they arise and are an effect of the fact that we are using a rotating coordinate system, rather than a so-called inertial frame of reference (i.e. we see the world sitting on the Earth and rotating with it, rather than looking at the Earth from space, which is a non-rotating frame of reference), and we still want to use Newton's second law of motion. This is a bit of a technicality which we will soon forget, and consider the above forces, just as normal forces.

Box 1   Meet the air parcel

The air parcel is a term used mainly to give an idea about what the forces act upon. It is very small, infinitesimally small in fact, and a way to think about it is as if a very small bubble of air has been coloured differently, so it is possible to follow it around as a result of the various forces that act on it.

We will now look at each of the forces in a bit more detail, beginning with the pressure gradient force:

$$\vec{P} = -\frac{1}{\rho}\vec{\nabla}p \tag{2.1}$$

where ρ and p are as defined above, and ∇ is the gradient, that is, the slope. The pressure gradient force is therefore defined as the gradient of the pressure field, that is, small differences in the pressure will result in a weak pressure gradient force, and large differences in a stronger force. Note the minus, which means that the flow will be pushed away from high-pressure areas and towards low pressures.

The frictional force, $\vec{F}$, is defined as follows

$$\vec{F} = -a\vec{V} \tag{2.2}$$

where a is the friction parameter, depending on the roughness of the underlying surface and the height above the surface. More rough surfaces and heights close to the ground will result in a high frictional force. The force is always directly opposite the wind vector.

The Coriolis force, $\vec{C}$, is defined as

$$\vec{C} = -f\vec{k} \times \vec{V} \tag{2.3}$$

where f is the Coriolis parameter ($= 2\Omega \sin\phi$, where ϕ is the latitude, and Ω the rotation of the Earth ($= 7.292 \cdot 10^{-5}$ rad/s)), $\vec{k}$ is a unit vector, pointing in the direction of the rotation vector (i.e. pointing straight out of the North Pole). The operator '$\times$' is the cross–product (remember the right-hand rule when figuring out the direction). In the Northern Hemisphere, the Coriolis force deflects the wind to the right, and in the Southern Hemisphere to the left. The Coriolis force is the most difficult of the forces to understand, but put simply, it has to do with the rotation of the Earth, and it works by turning the wind.

The gravitational force, $\vec{F}$, does not affect the horizontal motion, and is therefore often ignored when discussing the forces, but for completeness it is defined as

$$\vec{F}^* = G\frac{m_1 m_2}{r^2}\frac{\vec{r}}{r} \tag{2.4}$$

where F^* is the normal gravitational force, G is the gravitational constant ($= 6.674 10^{-11}$ N m^2 kg^{-2}), m_1 and m_2, two bodies pulling at each other (e.g. an air parcel and the Earth), $\vec{r}$ is the vector from the centre of mass of one body to the other, and r is the distance between the

Bio 1: Gaspard-Gustave de Coriolis[2]
1792–1843
France
French mathematician, mechanical engineer and scientist.

two. In the atmosphere we include an extra force, again due to the rotation of the Earth of the magnitude:

$$\vec{F} = G\frac{m_1 m_2}{r^2}\frac{\vec{r}}{r} + \Omega^2 \vec{R}_A \tag{2.5}$$

where $\vec{R}_A$ is the position vector which goes from the rotation axis of the Earth perpendicularly to the position of the body. The first part of this new force is therefore the normal gravitational force and the second an adjustment; so we can use it when doing calculations on the rotating Earth. As mentioned above, we use the gravitational force very little.

2.3.3 Force Balances and the Geostrophic Wind

We have now been introduced to the forces that govern the flow in the atmosphere. The three forces that we will focus on in the following are the pressure gradient force, the Coriolis force and the frictional force, since these are used to explain the so-called geostrophic balance and the resulting geostrophic wind. The term *geostrophic* stems from the two Greek words geo, earth-related, and strophe, a turning.

In the *free atmosphere* (defined as the part of the atmosphere where the frictional force can be neglected), we find – by definition – two of the three forces, the pressure gradient force and the Coriolis force, and when no acceleration takes place, these two forces balance each other out as illustrated in Figure 2.7. As can be seen from Equation 2.3, a certain strength of the force requires a certain wind speed. The wind speed that gives the Coriolis force that exactly balances the pressure gradient force is called the *geostrophic wind*, and the balance between the two forces is called the *geostrophic balance*.

Since we know the two forces, it is possible to write the two components of the geostrophic wind (u_G, v_G):

$$u_G = -\frac{1}{f\rho}\frac{\partial p}{\partial y} \tag{2.6}$$

$$v_G = \frac{1}{f\rho}\frac{\partial p}{\partial x} \tag{2.7}$$

---

[2] Figure source: "Gustave coriolis", https://commons.wikimedia.org/wiki/File:Gustave_coriolis.jpg#/media/File: Gustave_coriolis.jpg [Public domain], from Wikimedia Commons.

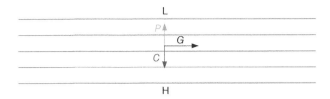

Figure 2.7   The two-component balance between the Coriolis force (C) and the pressure gradient force (P) found in the free atmosphere. The resulting geostrophic wind (G) is also shown. Horizontal lines are isobars (i.e. lines through points of equal pressure), and H and L indicate high and low pressure, respectively

Below the free atmosphere, that is in the Ekman layer, the friction force comes into play (getting stronger and stronger as we get nearer the surface, since the surface is the cause of the friction, through its roughness) and we now have a three-component balance, as illustrated in Figure 2.8: the pressure gradient, the Coriolis and the friction forces. Two things happen as the friction force is introduced: the wind speed is reduced, and the wind turns/veers (towards the low pressure).

It is possible to calculate the magnitude and direction of the geostrophic wind, and the expression for the speed, called the *geostrophic drag law*, is given below:

$$G = \frac{u_*}{\kappa} \sqrt{\left[\ln\left(\frac{u_*}{fz_0}\right) - A\right]^2 + B^2} \qquad (2.8)$$

where G is the magnitude of the geostrophic wind, u_* the friction velocity (see Chapter 4), $\kappa = 0.4$ the von Karman constant, f the Coriolis parameter, z_0 the roughness, $A = 1.8$ and $B = 4.5$ two constants (strictly speaking, the equation only holds when the atmosphere is neutral, see Section 4.8). This is one of the two basic equations in boundary-layer meteorology (the other is the logarithmic wind profile), and to try to explain what it means, focus first on G (the wind in the free atmosphere) and u_* which (as we will also see in Section 4.1) relates to the wind at the surface. So the equation says something about how the wind aloft relates to the wind at the surface. You might also recognise f, which means that the relation has to do with the latitude we are at (through the rotation at that location). Furthermore, the presence of z_0

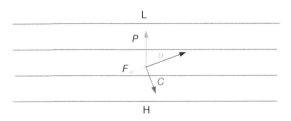

Figure 2.8   The three-component balance between the Coriolis force (C), the pressure gradient force (P) and the frictional force (F) found in the surface layer. The resulting wind (u) is also shown. Horizontal lines are isobars (i.e. lines through points of equal pressure), and H and L indicate high and low pressure, respectively

(which is related to the friction force) shows that the relation also has something to do with how rough the surface is (a forest will have a higher z_0, than, say, a field of grass, see Section 5.3).

The direction of the geostrophic wind can be found from

$$\sin \alpha = \frac{Bu_*}{\kappa G} \tag{2.9}$$

where α is the angle between the wind at the surface and the geostrophic wind, G, that is the turning of the wind. B, u_*, κ, G as above.

Very near the surface, that is, in the surface layer (see Figure 2.6), this three-component balance breaks down, and we are in what is called a well-mixed (with respect to among others momentum) layer, and in here the wind does not turn with height, and the vertical profile is given by the logarithmic wind profile (see Chapter 4 for much more on this).

One could ask oneself: how does the wind change as it moves from the top of the surface layer up to the free atmosphere. This can be solved theoretically and results in the beautiful mathematical shape called the *Ekman spiral* which is described further in Box 2.

Box 2   The Ekman spiral

It is possible to mathematically match the two-component balance with the three-component one. This gives the following expressions for how the x- and y-components of the wind (u,v) change with height:

$$u = u_G[1 - \exp(-\gamma z)\cos(\gamma z)]$$
$$v = u_G[\exp(-\gamma z)\sin(\gamma z)]$$

where u_G is the geostrophic wind, z the height above ground, $\gamma = \sqrt{\frac{f}{2K}}$, and K the eddy viscosity coefficient $\approx 1 \ m^2/s$. Plotting this, we get:

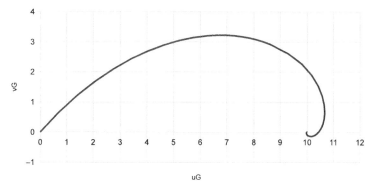

This is what is called the Ekman spiral, going from near the surface at the left of the plot to its final value of 10 m/s to the right.

Bio 2: Vagn Walfrid Ekman[3]
1874–1954
Sweden
Oceanographer. Studied physics at Uppsala University but became interested in oceanography during his studies. He discovered the Ekman spiral during an expedition on the good ship Fram with Fridtjof Nansen.

2.4   Length and Time Scales of Atmospheric Flow

Before we turn to an overview of the various weather systems typically found around the globe, we will pause and discuss the various scales (of length as well as time) one can find in the atmosphere. In many ways splitting the atmosphere into scales in this way is an artifact, of course, but as you will see, a very useful one.

In wind energy terms, the introduction of the various scales helps us think about the winds at our site all the way from the global to the very local level, and as such, helps us understand how a particular site is affected by a multitude of simultaneously working weather systems. This understanding is important when looking at the temporal and spatial variations found at the site, with respect to the diurnal (daily) variation, but also the seasonal (yearly) variation, and it also gives us a feeling for how long measurements should be taken and to what level of detail.

It is convenient to split the atmospheric phenomena into four scales: the global (or the planetary), the synoptic, the meso and then finally the micro scale. Typical dimensions of these scales can be found in Table 2.2.

Table 2.2   The four scales of atmospheric flow, the typical length and time scales are given. A characteristic phenomenon is also listed for each scale. O() means order of

Name	Length scale (m)	Time scale	Phenomenon
Global/planetary	10^7	O(weeks), 10^6 s	Planetary waves
Synoptic	10^6	O(days), 10^5 s	Cyclones
Meso	10^5	O(hours), 10^3 s	Sea-breezes
Micro	10^2	O(minutes to hours), 10^2 s	Thunder showers

---

[3] Figure source: "Ekman Vagn" by Unknown - [1], http://commons.wikimedia.org/wiki/File:Ekman_Vagn.jpg#/media/File:Ekman_Vagn.jpg [Public domain], from Wikimedia Commons.

As an example, imagine that you stand at your wind farm site, and you can see the nearby forests, hills and valleys, this is the micro scale, your site might be situated near a big lake or in a very wide valley, and the flow caused by such structures is said to happen on the meso scale. The synoptic scale flow is then the next scale up, and this is where we find low-pressure systems (powerful ones being storms), big high-pressure systems and so on. Finally, the location of your site on the globe, that is relative to the oceans, continents and the dominant mountain ranges, will determine the effect of the global or planetary scale; here we find for example, the jet streams (see Section 2.5.2.1). So, most of the atmospheric flow that actually affects your site is on scales way beyond what you can observe when just standing at the site, risking of course, that you do not take these flows into account in an appropriate manner when determining the wind resource at your site.

There is a natural upper limit to length scales on our (finite) planet, and that is usually given by the radius of the Earth (6371 km), and it is of course not possible for length scales to exceed this.

The various divisions of atmospheric flow described here do not only have a typical length scale associated with them; there is also a typical time scale. This is a little bit more difficult to understand, but imagine a time scale as the typical time it takes for a phenomenon to go through its various stages. As an example, consider a low-pressure system which is formed, grows and then disappears in a matter of days. This, combined with the typical length scale of low-pressure systems, means that they are on the synoptic scale. Similarly, the time scale of small thunder showers (on the micro scale), is found to be minutes to hours. The remaining time scales can be found in the table. One could add another two, one at the smallest scale, turbulence (length: mm/cm, time: seconds) and one on the very largest, the climate change scale (length: still the radius of the Earth, time: tens of years), but for our purposes the four described in the table are the most important ones.

As mentioned at the beginning of this section, these scales are somewhat arbitrary, but seeing the flow in this way, at these different scales, does help the understanding of the flows encountered, and it provides a good framework for analysing observations, and as you can see, they do split the various types of flow into quite useful categories. As a simple illustration of this, see Figure 2.9.

Figure 2.9 A hand-drawn sketch of a time series of wind speed, attempting to illustrate how the different scales can be seen in measurements at a site. This is intentionally left sketchy in order to avoid any resemblance to a real time series, because it is never so simple in real life. In the plot, you can see the long-term average, the straight line, but also that there is significant variation around this line. You can also see the diurnal cycle and the signature of a storm

It is also important to stress that the scales, of course, are interlinked. We have already seen the connection between the geostrophic wind (which can be considered on the meso to synoptic scale) and the local wind at the surface (the micro scale), through the friction velocity (u_* in Equation 2.8). In fact, there is a general tendency for the larger scales to drive the flow on the smaller scales.

As you can see from Table 2.2, once the various phenomena found in the atmosphere have been classified, there is a clear tendency for large phenomena to have long time scales, and smaller phenomena to have shorter time scales. We will see some deviations from this later in this chapter, but, again hand-waving, one can imagine that very large phenomena have a lot of inertia (due to their size) and by definition cover large areas and therefore it takes time for the system both to grow but also to evolve. The opposite goes for the small phenomena.

2.5   Larger-Scale Systems (aka Weather)

Having discussed the various scales that can be found in the atmosphere, we will now turn to some typical examples of what I guess can best be described as 'weather'. In the following I will cover the four main ones, and they are:

- mid-latitudinal cyclones (low-pressure systems),
- anticyclones (high-pressure systems)
- hurricanes/typhoons/severe tropical cyclones and
- monsoons.

Using our scale glasses from the previous section, it can be seen that we are looking at systems at different scales: severe cyclones are quite big (a few thousands of kilometres), but fairly short lived, and as such do not completely fall into the categories in the previous section; however, the mid-latitudinal cyclone and the anticyclone are the archetypes of synoptic scale systems, and the monsoon is an example of something which length-scale wise is between the synoptic and the global scale and time-wise actually months.

It is important to stress again that the atmosphere is a chaotic three-dimensional system, and as such all the typical weather phenomena cannot be described in simple terms. So when we attempt to do so, as in the following, it is necessary to cut some corners, in order to come up with simpler images of how things work.

2.5.1   Mid-Latitudinal Cyclone (Low-Pressure System)

The low-pressure systems are very closely connected to the so-called fronts. I am not entirely sure if the following is a true story and where it comes from, but I really like it, so here we go! In the early days of the field of meteorology, the groups of society that had most interest in the weather were the farmers (but they did not have either resources nor organisation to do much else than to worry about it!) and the military (they had plenty of both). So once systematic measurements of mainly temperature and pressure were carried out and people realised that there were actually big areas of air that were distinctly different from each other, and they were right next to each other, too, for example with a very sharp boundary with respect to temperature, it seemed very obvious to use a military term for the places where these warm and cold air masses met and 'fought' against each other, *fronts*, and so, the terminology of fronts was born. As I said, not sure this is how it really happened, but I think it is a nice story anyway.

So, in the places where the warm air masses are attacking, or moving in on the cold air masses, we have a *warm front*, and where the cold air masses move in on the warm air, we have a *cold front*. In yet another example of how complicated and three-dimensional the atmosphere is, it is quite difficult to describe how the fronts are created in the first place. This is called *frontogenesis*, Greek for the creation of a front. To illustrate how complicated this is, elements that contribute to the creation of fronts are described in for example, Hoskins and Bretherton (1972) where they identified as much as eight different mechanisms that influence the development of these temperature gradients: horizontal deformation, horizontal shearing, vertical deformation, differential vertical motion, latent heat release, surface friction, turbulence and mixing, and radiation. In more technical language what happens is the formation of a *baroclinic* wave (see Box 3).

Box 3 Baroclinic and barotropic atmospheres

The definition of a baroclinic atmosphere is one where the density depends on both the temperature and the pressure. Its 'opposite' is a barotropic atmosphere, where the density depends only on the pressure. The barotropic areas on the Earth are generally found in the tropics, and the baroclinic areas are generally found in the mid-latitudinal/polar regions.

The way the fronts then develop into a low-pressure system is called *cyclogenesis*, Greek for the creation of a cyclone, that is a low-pressure system, and the development is as follows (using the original (simple) theory of the so-called Bergen school (Bjerknes, 1900), c.f. Figure 2.10).

Two air masses, a warm and a cold, meet. A small disturbance (a baroclinic instability) starts to develop (Figure 2.10a). A wave starts to form where the warm air starts to move in on the cold air (at the warm front) and the cold on the warm air (the cold front) (Figure 2.10b). The wave continues to develop. At some point, the cold front starts to catch up with the warm front and what is called an *occluded front* starts to develop (Figure 2.10c). Finally, the cyclone is fully developed (Figure 2.10d).

Where the warm and the cold fronts meet is where we find the low pressure, and as the low-pressure system develops the pressure gets lower and lower. As we know from above, the lower the pressure (gradient), the stronger the winds. A good way of remembering which way the wind blows is the LLL rule: 'leaving low to the left', that is, if you sit on an air parcel you will move with the low pressure on your left (on the Northern Hemisphere and on your right on the Southern).

Furthermore, we find clouds and rain (precipitation is the posh term for rain, snow, etc.) around the fronts, most pronounced at the warm front and the occlusion.

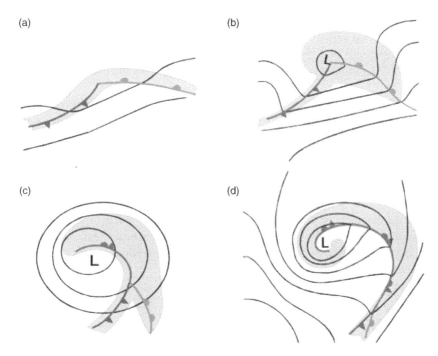

Figure 2.10 The development of a low-pressure system from a small wave to the fully developed low-pressure system with warm (semicircles), cold (triangles) and occluded fronts (semicircles and triangles). See text for an explanation of the various stages. Solid black lines are isobars, L denotes the low pressure, and the grey areas are clouds, often associated with rain. Source: © EUMeTrain. Reproduced with permission of EUMeTrain

After 2–5 days the fronts start to dissolve (called *frontolysis*) and the low pressure starts to 'fill up' (i.e. the pressure increases again), and the cyclone ceases to exist. As mentioned above, this is a very simple description of how a low-pressure system develops. But if you look at movies of satellite pictures, you will see that the overall system very often follows this development, but of course also with some weird and wonderful variations! As you can see from the movies, not only does the low-pressure develop as described above, it also moves east-wards at the same time, as if on a conveyor belt – so yet another example of the fact that the atmosphere is a complex, complicated, three-dimensional system!

2.5.2   Anticyclones (High-Pressure Systems)

Since we have just discussed the low-pressure system (the cyclone), it would make sense also to briefly mention its counterpart, the high-pressure system (or the anticyclone). The high-pressure system is not as 'famous' as the low-pressure system, since it is not in the same way associated with bad weather (rain and storms). Actually it is quite often associated with rather good weather. However, it is worth just spending a few lines in the company of the high-pressure system. Just like it was the case for low pressure, a high-pressure system is found where there is a local maximum in the pressure field, and the isobars are closed around this area. Since the pressure is high relative to its surroundings, the air at the surface will flow away from the high pressure, which means that the air above the area will sink. Over land, the high-pressure area will generally be dry and free of cloud (due to the descending dry air).

The wind spirals outwards from the high-pressure area, and opposite (hence the anti- in anticyclone) to the low pressure, that is clockwise (on the Northern Hemisphere). The winds are generally quite calm and the high pressure normally covers quite a large area, and almost always larger than the low pressure. High-pressure systems will often live longer and move less than a low-pressure system.

2.5.2.1 Jet streams

Looking at the atmosphere vertically where the warm and the cold air masses meet, at about the height of the tropopause (i.e. at 5–15 km's height, see Table 2.1), we find the *jet stream*. This is an area with very high winds, up to 90 m/s, over very wide areas, up to 300–500 km, and the jet stream itself can be many thousands of km long. The winds are generated due to the very pronounced difference between temperature and pressure, combined with the fact that the Earth is rotating (remember Coriolis). We find the following three types (c.f. Figure 2.11):

1. The *polar jet* with predominantly westerly winds, and closely connected to the low-pressure systems described above, found on the Northern as well as on the Southern Hemisphere.
2. The *subtropical jet* with predominantly westerly winds, also found on both hemispheres, but at significantly higher altitudes.
3. The *tropical jet*, which, as opposed to the two other types, have predominantly easterly winds, this jet is over the Equator, so there is only one. These jets are not so dominant, weaker and more rare (not shown in Figure 2.11).

In some sense (it is difficult really to tell who is the chicken and who the egg here) the jet streams 'pull' the weather from west to east especially the low-pressure systems. Another characteristic of the jet streams is that they meander (i.e. wiggle), and when this happens they can also help in forming the low pressures (in the troughs) and highs (in the ridges). Anybody

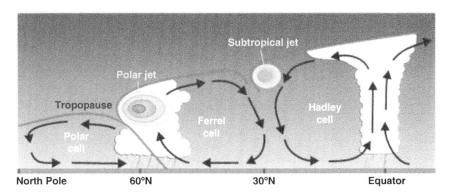

Figure 2.11 Illustration of the two types of jet streams (as mentioned in the text, the tropical jet is weaker and not always present). Note again that it is the jets which are the important feature (and not the three cells) and that they blow in/out of the paper. Source: http://www.srh.noaa.gov/jetstream//global/images/jetstream3.jpg

who has flown intercontinentally is very familiar with these jet streams, since the pilots hunt them, in order to fly significantly faster than usual (going with the jet stream), and avoid them (going in the opposite direction) in order not to be slowed down.

2.5.3 Hurricanes

A famous, or rather infamous, and very spectacular weather system is the *severe tropical cyclone*, better known as the hurricane around North America and the typhoon in Asia. A cyclone becomes – by definition – a severe cyclone/hurricane/typhoon (I will use hurricane in the following) when the wind speed is above 75 mph (33.5 m/s). These systems depart from the usual relation between time and length scales (see Section 2.4) in that they are fast, but also rather large (up to 4000 km). Hurricanes are formed only over oceans, only if the sea surface temperature (SST) is above 26.5°C, and only at latitudes higher than 5° North or South. Due to these constraints they are found in the Caribbean, the Pacific and the Bay of Bengal. The major 'fuel' of the hurricane is the latent heat of the moist air, which explains why the temperature has to be high. The Coriolis force (Equation 2.3) is the other driving force and, as we know, it gets smaller and smaller as we get nearer the Equator, which helps explain the 'band' around the Equator.

A characteristic of the hurricane is its *eye*, in this strange place the winds are calm, the air dry and sinking (rather than raising, which is the case in most of the hurricane). The eye is often very well defined with a clear 'wall' surrounding it, the radius is typically around 15–30 km, a beautiful example of a hurricane can be seen in Figure 2.12.

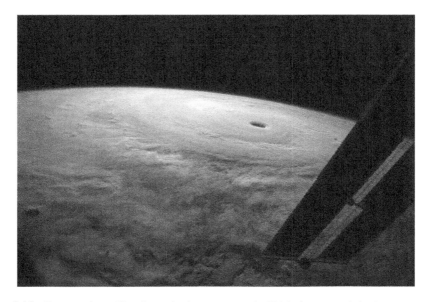

Figure 2.12 Super typhoon Vongfong, the largest storm in 2014, the eye and the immense scale is clearly seen. Taken from the International Space Station by astronaut Reid Wiseman who posted it on his Twitter account from space! Source: NASA/Reid Wiseman

> **Box 4** What makes a hurricane/typhoon?
>
> Empirical evidence has shown that the following three conditions need to be met in order for a hurricane to develop:
>
> - Over water
> - Sea surface temperature greater than 26.5°C
> - A location with a latitude higher than 5°(north or south)

As you can imagine – due to their devastating nature – a lot of research is going into understanding and modelling these hurricanes; however, as you probably have witnessed on the television and news, we still do not understand them well enough to accurately predict where they make landfall, nor when they die out, which they inevitably will when over land, due to the lack of the moist air, their fuel.

2.5.4   Monsoons

The term 'monsoon' originates from the Arabic word 'mawsim', which basically means 'season', and is used to describe the seasonal changes in wind speed and direction and precipitation. In the traditional definition of Monsoon, only the West African and Asia-Australian monsoons are included. The monsoons are mainly known for the rainfall associated with them, which is not all that strange since for example, about 80% of the rainfall in India happens during the wet part of the monsoon season (Turner, 2013). However, in the case of wind energy it is equally important to know that the wind climates of the two seasons (monsoon and non-monsoon) are very different, and treating data from a monsoon area without taking this into account can easily lead to wrong conclusions (e.g. when looking at the distribution of the winds at a site, see Section 3.3). The two seasons need to be identified and separated. The monsoon will have different effects on the wind in different areas. In general, the summer monsoon will bring westerly winds and the winter monsoon (non-monsoon) easterly ones. Also, the monsoons are quite irregular, so again when looking at data from monsoon-affected areas, this needs to be taken into account.

The monsoons are caused by a very large sea breeze (see Section 5.5) formed (taking the Indian monsoon as an example) by the difference in heating between the Indian Ocean and the Indian subcontinent. The onset of the monsoons is caused by the movement of the ITCZ (see Section 2.1) as the seasons change.

2.5.5   Climatological Circulations

At the end of this chapter on the basics of meteorology, I wanted to describe two phenomena that span large areas and long time intervals. The first is the El Niño, the second the North Atlantic Oscillation (NAO). The El Niño can be found in the Pacific and the NAO in the Atlantic.

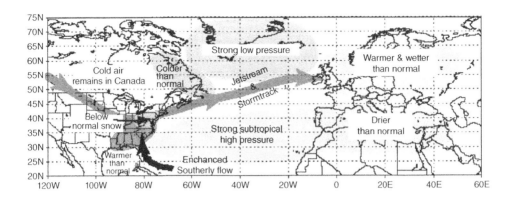

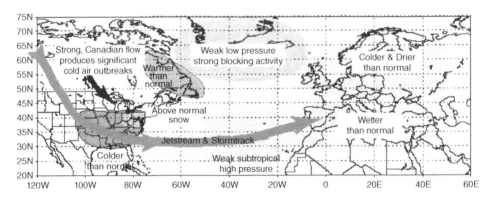

Figure 2.14 The two phases of the NAO: the positive phase (top panel) and the negative phase (lower panel). The effects on the weather, winds and precipitation are described. Source: NOAA

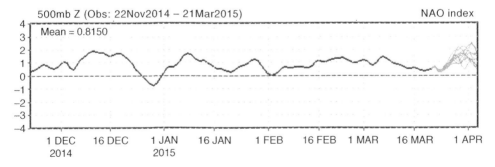

Figure 2.15 Recent values of the NAO index, and an ensemble forecast for the future values, see Chapter 8 for more on ensemble forecasts. Source: NOAA

The NAO was discovered by Sir Gilbert Walker in the 1920s, and is an example of a *teleconnection*, that is a phenomenon where the effects are related/connected over very long distances (thousands of kilometres). There is some evidence that there is a correlation with wind speed in Northern Europe (and other places) and the NAO index (e.g. Hodgetts (2011) found good correlation during the winter months between United Kingdom wind speeds and the NAO index).

Bio 3: Sir Gilbert Thomas Walker[4]
1868–1958
United Kingdom
Mathematician and meteorologist, who became Director General of Observatories in the Indian Meteorological Service. He also identified the Southern Oscillation, see section above on El Niño (2.5.5.1).

2.6 Summary

This chapter has described the basics of general meteorology. We started with the very fundamental question of why the wind blows. We then went into a description of the different vertical levels of the atmosphere, starting with the troposphere going all the way up to the exosphere. Zooming in on the troposphere, it could be subdivided into a number of layers. Nearest the surface we find the surface layer, above this, the Ekman layer and last the free atmosphere.

The atmospheric forces and variables were then described, and we looked at pressure, temperature, density, velocity and humidity. Knowing these enables us to describe the atmospheric forces, that is the forces that work on an air parcel. The four forces we described where: the pressure gradient force, the fictional force, the Coriolis force, and the gravitational force. Of these, the gravitational force will be used the least in the following.

The geostrophic wind and the two- and three-component force balances were then introduced. In the free atmosphere, that is, where there is no effect of the frictional force there is a two-component balance between the pressure gradient force and the Coriolis force. As we get nearer the surface, the frictional force starts to have an effect, resulting in a three-component balance between the forces. Matching these two force balances, it was possible to derive the first of two very important equations, the geostrophic drag law. The geostrophic drag law connected the wind aloft that is, in the free atmosphere to the wind at the surface.

Going up through the Ekman layer, the wind speed increases and the direction turns, described by the Ekman spiral.

Slightly on the philosophical side, we then had a look at the various length and time scales that are connected with the different types of atmospheric flow. There was a general tendency of having short length scales connected to short time scales and similarly long length scales connected to long time scales. Philosophy aside, using the scales as a way of thinking about the flow at your site, is quite a useful way of understanding all the effects that can govern the flow there.

The last part of the chapter described four large-scale weather systems what we would normally just call weather. The four systems were the mid-latitudinal cyclone (low-pressure system), the anticyclone (high-pressure system), hurricanes and monsoons. When talking about the mid-latitudinal cyclone we introduced the term front, of which there is a warm as well as a cold front.

In connection with the low-pressure systems and somehow connected to the development of the low-pressure systems, we also briefly described the jet streams of which there are mainly three. These are the concentrated areas high aloft in the atmosphere where the wind speeds are very high.

High-pressure systems (or anticyclones) are in many ways the poor cousins of the low-pressure system. They are characterised by low winds, and fair and dry weather.

Next, hurricanes were described. These are also called, more correctly, severe tropical cyclones, but depending on the region also known under the name typhoon. These are weather phenomena driven by some very specific atmospheric and oceanographic conditions. Hurricanes are mainly found in the Caribbean, the Pacific and the Bay of Bengal.

The last weather phenomenon was the monsoons. In general terms, these are caused by very large scale sea breezes where air rises over big land masses (such as the Indian subcontinent) and sinks over the oceans (such as the Indian Ocean). Where they are present, they have a significant influence on the weather all year round.

At the end of the chapter we described two phenomena that span large time scales and cover large areas: El Niño and the NAO. The former is found in the Pacific and the latter in the North Atlantic. Both of these phenomena are described by an index in the case of the El Niño, the SOI and in the case of the NAO, the NAO index, it is the sign and magnitude of these that define the phenomenon.

Exercises

2.2 *A warm-up question: what does Newton's second law of motion state?*

2.3 *Calculate Ω, the rotation of the Earth.*

2.4 *What is the magnitude of the Coriolis parameter at the Equator? On the North Pole? On the South Pole? Where you live?*

2.5 *You have two straight and parallel isobars: one is 1000 hPa and the other 1010 hPa, they are 500 m apart. Calculate the x- and y-component of the pressure force.*

2.6 *Calculate the geostrophic wind speed and turning given: $u_* = 0.3$ m/s, latitude $40°N$, $z_0 = 0.1$ m.*

2.7 *Calculate the friction velocity, given a geostrophic wind speed of 8 m/s, latitude $55°S$, $z_0 = 0.05$ m.*

2.8 *Why does the atmosphere stick to the surface even though the Earth rotates very fast in space?*

2.9 *Find the value of the Ekman spiral at 300 and 1000 m. Assume a geostrophic wind of 9 m/s, and a latitude of $30°N$.*

2.10 *Identify the various phenomena and scales that affect the flow at your current location.*

2.11 *Find a satellite picture (ideally a video) of a low-pressure system from your local met office/institute. Try to determine the length scale, and if possible also the time scale of the system.*

2.12 *Looking at the same picture/video try to find the warm and cold air, the fronts, and, as the system develops, identify the occlusion, and try to see how the whole system weakens and finally disappears.*

2.13 *Looking at the same picture/video try to trace a front and determine how quickly it moves.*

2.14 *Does the air rise or sink over a low pressure?*

2.15 *Try to locate a high-pressure system on a weather map, and if possible see how fast it moves and how big an area it covers.*

2.16 *Search the web for the latest science concerning El Niño.*

2.17 *Are we in an El Niño or a La Niña year, currently?*

2.18 *What is the NAO index currently?*

2.19 *Find the names of two other teleconnection phenomena.*

3

Measurements

You arrive at the site early in the morning, after a night at a basic but nice hotel in the middle of nowhere. Once you have oriented yourself, the questions quickly start to arise: where should I place my masts/measuring devices; which kind of instruments should I use; what is the purpose of the measurements; how accurate are they, and many other questions like that. In this chapter, I will try to cover these basics, some philosophy and a bit about how to treat the resulting measurements. I will also go through a long list of instruments that we use in wind energy, explain what they do and how they work. First the philosophical part with some generic and general points and discussions.

3.1 Philosophy: What Does it Mean to Measure?

We measure at a site to obtain knowledge about the quantities we are interested in at that location (e.g. the mean wind speed). In many ways we consider the measurements as fact, ground truth, how it is, because having been at the site, carried out actual observations of what we are interested in gives us a feeling of knowing. A simplistic view would then be that 'we measure to know'. The remaining sections of this chapter will qualify this statement, qualify it quite a bit, actually!

To frame the discussion we will look at three terms: representativity, resolution and accuracy. The two latter terms are quite interrelated, but splitting them helps to highlight the important points. Accuracy and resolution have also a strong relation to precision, but the terms are not the same, so I will also briefly discuss that difference.

Turning first to representativity, one thing is to have been at the site and having done *some* measurements, a completely different is to determine if the measurements answer the question we would really like to have answered. These questions can naturally be many, but a typical question in a wind energy context could be: how many MWh will my wind farm produce per year for the lifetime of the wind turbines? To answer this question we

Meteorology for Wind Energy: An Introduction, First Edition. Lars Landberg.
© 2016 John Wiley & Sons, Ltd. Published 2016 by John Wiley & Sons, Ltd.

will need to break it down into quite a few very specific subquestions, like what is the climatologically averaged sector-wise distribution of the wind speed? This is by far the most important question to ask, but if for example, stability (see Section 4.8) plays a big role, then measurements of temperature differences or heat fluxes (see later) will also play an important role.

Using the question about the sector-wise distribution as an example of what representativity means, a further two important questions arise: *where* to measure (in a three-dimensional sense) and for *how long* in order for the measurements to be representative? Taking the 'where' first: imagine a wind-farm site many kilometres wide and long, in a varying terrain with hills, valleys, grass land and forests, and perhaps a lake somewhere as well. Trying to find out exactly where to put the mast, and probably more importantly, how many masts to put up, depends on a long list of things, as you can imagine, which I will not go into much further here, but the list includes proximity to the planned turbine locations, making sure that the various likely different physical flow patterns are represented in the measurements, etc. Locating the mast(s) on the site is the first part of the 'where' question, the second is at which height? Again, as I am sure you will be happy to know (!), this depends also on a long list of things, the two important ones are: the hub height of the wind turbines you plan to erect and if there are any characteristics of the site that would require a certain minimum height, like the aforementioned forest. As we go through the chapters of this book, you will be able to inform the decision about where, how many and at which height to quite an expert level.

Having now covered the 'where', we will need to address the 'how long', which is also a complex question to answer.

Exercise 3.1   *How long should you measure for to obtain values (of say the wind speed) that are representative for estimating the production over the life-time of a wind farm?*

- *a few seconds*
- *a minute*
- *an hour*
- *a day*
- *...*
- *a year*
- *a few years*
- *20 years (the typical life time of a wind turbine)?*

Solving the exercise here (but please consider it first yourself, too), I guess ideally speaking measuring for the life-time would be the best, or to be more precise, measuring for as long as required for the average wind speed to be *climatologically representative*. Climatologically representative means that the average that you have arrived at is the long term average, that is, you will have seen all the typical variation during the measuring period. Practically that is of course quite silly, since you would like to put up the wind farm now or at least as soon as possible! Clearly, a few seconds to a few days are too short, since just looking out the window at how the weather develops and changes, quite different winds can be experienced over a few days, and also weeks. We have talked about time scales in Section 2.4, and in most places of the world, the winds we see over the seasons do also vary, so measuring for at least a year will cover those. As we saw in the previous chapter, the annual winds can also vary from year to year. We are now at the stage where trade-offs will be required (in order not to end up measuring for the entire life-time of the wind farm), as we will see in Chapter 8, computer-generated very long time series of wind speed and direction of quite high accuracy and representativity do exist and there are ways to link measurements on site to these time series to come up with long synthetic on-site data sets of good quality (see the description of the so-called MCP method in Box 25). In the old days (i.e. not that long ago!), before the high-quality computer generated data, the only method of finding appropriate long-term references was to use nearby observations. These were typically taken from airports, light houses, and meteorological stations run by national meteorological services. These data sets are still in use, and the method used is exactly as described above.

We have now discussed the first of the topics, representativity, and will turn to the second: *resolution*. In order to have observations/measurements that are of any use, they of course need to be of such a resolution that they actually resolve what is being observed down to a detail that will give the information we are looking for. As an example, in a wind energy context, think again about the measurement of the on-site wind speed. Here we need to have enough resolution to be able to determine if the project we are planning to build is financially viable. Given the detail required by financial models, in many cases this would demand a resolution of 0.1 m/s, so we need to have instruments that are able to resolve the wind speed to that resolution or better.

The last term we will discuss is *accuracy*. An instrument might be able to resolve a quantity to a certain level, but if it is not accurate, then it is of little use, so when reading from an instrument, it is not enough just to see if the instrument has the appropriate resolution but one also needs to see if it is accurate. As an example, imagine that an instrument has a resolution down to 0.1 but the accuracy is only 0.2, which means that we have a number which – for the wanted purpose – is not appropriate, since a measurement of for example, 3.1 could mean 2.9 as well as 3.3. Note that the determination of an instrument's accuracy needs external estimation. Again using wind speed as an example, the cup anemometer (see Section 3.4.2) needs to have its accuracy determined by carrying out measurements in a controlled environment like a wind tunnel.

As mentioned above, there is a term closely related to resolution and accuracy and that is *precision*. As you can see from the definition in Box 5 below, this is often thought of as relating to the number of digits. This can sometimes be quite misleading, because precision is not the same as resolution, as we will see in the following exercise:

Exercise 3.2   *Determine the precision and resolution of the instrument that has produced the following output: 26.068, 36.708, 29.792, 3.192, 17.024, 43.624, 3.724, 40.432, 43.624, 22.876, 5.320, 20.748, 28.196, 25.536, 48.944, 32.452, 45.752, 1.596.*

Did you think the precision and the resolution were the same (0.002)? Then please go back and look at the numbers again carefully (the answer can be found in Appendix B).

Box 5   Dictionary definitions of terms

Representative: serving as a portrayal or symbol of something.

Resolution: the smallest interval measurable by a scientific (especially optical) instrument.

Accuracy: the degree to which the result of a measurement, calculation, or specification conforms to the correct value or a standard.

Precision: refinement in a measurement, calculation or specification, especially as represented by the number of digits given.

We have now looked at some of the philosophical/theoretical aspects of measuring and will turn to a discussion of what to do once we have the measurements recorded in some way. But before we do that a good way to end the discussion about representativity, resolution, and accuracy, is to try to answer the question: when are two measurement values the same and when are they different? I shall illustrate what the point is in two different ways as an example, by comparing 0.8 and 0.8 and then 21.4 and 22.8. This might sound very cryptic, which is also the point! Let us take 0.8 and 0.8 first, are these two numbers the same? At first glance, the answer is, yes, of course they are, but what if the actual observed quantities that resulted in the two values were 0.8143 and 0.7732? If the precision of the instrument is only 0.1 then we would not be able to distinguish the two numbers, but what if the resolution was 0.0001 and the precision of the instrument allowed for that to be recorded, then the two numbers actually end up being quite different, especially if the resolution we need for the task at hand also is 0.0001. So, depending on the resolution as well as the precision of the instrument, the two numbers could either be identical or very different!

Turning now to the second pair, 21.4 and 22.8, are these two numbers identical? Given the precision and assuming a resolution of the instrument that looks much smaller than 1.0, we would say that no, the two numbers are not identical. But, we have forgotten to think about the *accuracy* of the measurements! If the accuracy is say, 1.0 or maybe even more, it becomes impossible to distinguish between the two numbers, and we are not in a position to say that they are not identical; however, if the accuracy is, say 0.01, the two numbers again become different.

I hope this playing with some very simple numbers has started you to think about data, precision, accuracy, resolution and finally representativity in a new way. Also that it will help you when you have to report a number, for example, by not including too many or too few digits of the number in question.

Box 6 Bias and scatter

There are two further aspects to the discussion of accuracy and that is bias and scatter/spread. The two numbers say something about two aspects of the data: how close are we on average to the value we are looking for (bias) and what is the spread of the data (scatter)?

In more mathematical terms, the bias is the mean of the data, and the scatter the standard deviation.

It is important to distinguish between these two numbers, so below you can see four cases (the dots are the data points, spread out horizontally in each group to make the scatter clearer):

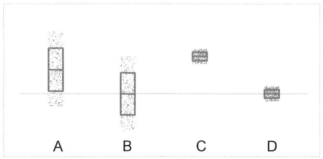

In case A, we have a large bias. The data points are on average far away from the line (which is to be considered as the goal), and we have a large scatter too. In case B, the data points are, on average, very near the goal, but the scatter is still large. In case C, the bias is quite large, but the scatter is small, whereas in case D, the bias is very small and so is the scatter.

I have put some lines and boxes on the plot too: the box is plus and minus one standard deviation, and the horizontal line inside the box is the mean (i.e. the bias).

It is very common, when talking about bias and scatter to represent the distribution of all of the data with a Gaussian distribution, it is surprising how often this distribution is a good fit to the data. The Gaussian distribution is shown below (where μ is the mean, and σ the standard deviation).

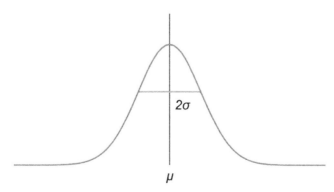

Please see Exercise 3.11 for a bit more on this distribution.

3.2   What Do We Measure?

Most of the data we meet is given to us in the form of a digital file and has the following format:

Time1 value1
Time2 value2
Etc.

Turning to our good old friend, the wind speed, the data could look as follows:

201501150930 8.3
201501150940 6.5
201501150950 7.1

Which, as you have already figured out, I am sure, represents the time (the first number): year (2015), month (01) for January, day (15), hour on a 24-hour scale (9 A.M.), minute (from 30 to 50), and then the wind speed in m/s (the second number).

But what do those values actually mean? And what happens to the wind speed during the 10 minutes between the data points? The answer to this question depends very much on the instrument in question and the way the data is treated after the measurement has been carried out. Some instruments record the instantaneous value measured at the given time, but most do some kind of *averaging*. Before we discuss this any further, let us pause for a second and describe the mechanics of getting the signal from the instrument to the digital file we saw above. In most cases, the instrument is connected to a device called a data logger. The purpose of the logger is to treat and store some or all of the data that the instrument has measured. The raw data that comes out of the instrument is often an analog electric current and that needs to be transformed in some way to convert it into a digital value which can be stored in a file. These loggers can be very simple, as is the case at a very remote site with no connection to the outside world, communication and power-wise, where only the most important numbers (e.g. the 10-minute averages) are kept. But they can also be very complicated, as with a system with lots of masts and instruments, with lots of processing power directly connected to the internet, enabling anyone, with the right credentials, to access the data from anywhere.

Returning to the averaging, there is nothing set in stone about what the correct way of doing this is, so the first and most important rule is to find out how it has been done in the case of the data you are looking at and then consider if it will have any effect on the results seen in the light of the purpose the data is to be used for.

There is one standard that is and has been used for many wind speed data sets and that is the one of the WMO (the World Meteorological Organisation under the UN). This standard says that wind speed should be averaged over 10 minutes and reported every hour or every 10 minutes (WMO, 2008). Which is probably why we see a lot of wind speed data averaged over 10 minutes.[1]

---

[1] In the case of wind speed there is also a good argument based on the various time scales found in the atmosphere, but I will not go into that in any details here.

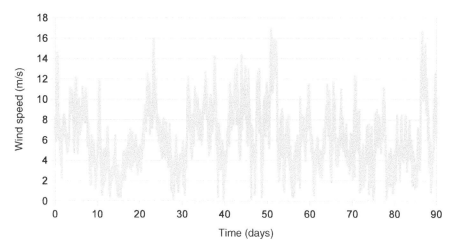

Figure 3.1 The speed (*y*-axis) and time (in days, *x*-axis) of the example time series discussed in the text

We now have a basic understanding of how the various instantaneous measurements come from the instrument to some kind of digital representation. In the following, we will start looking at different ways the data can be treated and analysed and for that we will need some data which we can study in a number of exercises, a small sample of the data is listed in Appendix C, and the full set can be downloaded from the website under Data (QR code in the margin).[2]

3.3 Measurement Theory, or Different Ways to Treat the Data

In this section, we will discuss the basics of what could be called measurement theory, that is, how to look at and analyse data in general and measurements in particular. This is a very big field, and, as with the other topics in this book, I will only scratch the surface, but still try to cover the basics.

The first thing you need to do when you get some new data is to get a first idea about it. Normally this is done through some kind of visualisation, most often a two-dimensional plot of the signal versus time, but plotting several signals in the same plot can also be very illustrative and reveal relations that would otherwise have been very difficult to spot. As an example, I have plotted our example wind speed data in Figure 3.1.

As you can see, we have almost 90 days of data, equal to 3 months. As discussed above, this is of course nowhere near the length of time we need in order to use it for something practical, it is only so we have some numbers to work with. If you look at the file, you will see that aside from a time stamp and a wind speed value, there is also a third column, and

---

[2] Permission to use the measurements for these exercises has kindly been granted by the copyright owners of the internet database: 'Database of Wind Characteristics' located at DTU, Denmark. internet: http://www.winddata.com/.

that is the wind direction (which we will use later in this section). The minimum value is very near 0 m/s and the maximum around 17 m/s. It might sound like a funny remark, but the data also *looks* like wind data! You will see that, as you analyse more and more wind speed (and direction) data, you will develop a feel for what it should look like, and if you compare for example a plot of share prices, you will find that the 'nature' of the plots is very different. There are three high-wind cases in the sample, one just after Day 20, another just after Day 50 and finally one just before Day 90. Of the three cases, the first one (the one after Day 20) is the one that resembles a 'normal' storm (i.e. the passing of a low-pressure system as discussed in Section 2.5.1) the most, with the slow increase and gradual decrease after the peak of the event.

It can sometimes be hard to see from a visualisation like the one above, but identifying *missing data* is an important task at this stage, too. Missing data means that for some reason or another (say power failure, a mechanical problem, lightning strike, human interference (!)) a value of the quantity (wind speed, direction, temperature, etc.) in question has not been recorded at a particular time. Typically a specific value is used to signal that the data is missing. For our data set it could look like:

201501151950 99.9 999

Indicating that both wind speed (99.9) and direction (999) are missing for this time and date.

If there is a lot of missing data, and in particular if the data is not missing at random, that is there is a pattern, it is very important to identify this, before any further analysis is carried out (think for instance of including a value of 99.9 when calculating the mean wind speed!). There can be many reasons for patterns in the missing data; my favourite example is from a long time back when I analysed data from a meteorological station in a country I will not mention the name of, the specific data set I was looking at was read off manually, and we realised that there was a very clear pattern in the missing data: no measurements at night during the winter! Investigating this, we found that it was too cold for the person to go out and read the instruments (at least according to the person)! This, as with many other patterns, will of course bias the measurements, making them non-representative and impossible to use directly. Furthermore, if some kind of data interpolation/padding is used, increase the uncertainty quite a bit, too.

Depending on the way the data has been recorded and treated, there might also be another, and the most obvious, type of missing data: where data is actually really missing from the file, that is, where data from one or several time steps are not recorded, and the time stamp cannot be found in the file. Depending on how you plot this, this can actually also be difficult to spot, especially if it is only one or a few data points, and any patterns in this can easily be missed. Longer periods of missing data will of course be easier to spot.

Exercise 3.3   *If you plot a time series of some data versus time, how would missing data look?*

One indication of how good the quality of the data is, is through the *recovery rate*, R, which is defined as the ratio between the number of valid data points actually in the file divided by the number of data points that could have been in the file, had we recorded one data point for each time step for the entire duration, viz:

$$R = \frac{\text{No of valid data points}}{\text{No of possible data points}} \tag{3.1}$$

The recovery rate can be used as a coarse measure of the quality of the data, but it of course does not reveal the patterns we discussed above. Most people would say that R should be above 90% to have data of a sufficient quality for wind energy purposes.

Once you have a good idea of the data in a visual sense and have an understanding of the missing data, you need to calculate some statistics relating to the data. As a minimum, the mean/average and the standard deviation are two good numbers to have (see Box 6). In the case of the above data, we get (please put the data into a spreadsheet and calculate the numbers yourself before you proceed):

Mean wind speed: 6.2 m/s
Standard deviation: 3.0 m/s

In Figure 3.2, I have plotted the wind speed data again with the mean and standard deviation overlaid, so you can get a feeling of what the two numbers mean. If you are unfamiliar with these terms, then the mean/average is the 'middle' value (the sum of all the values, divided by the number of values) and the standard deviation is a measure of the scatter, that is, how much the data spreads out around the mean. If you take the interval between plus and minus one standard deviation you will have covered 68.2% of the data. Please refer back to Box 6 for more on this.

Up until now we have discussed wind speed a lot, but it is not until wind speed is linked to wind direction that the data really starts to make sense. We not only need to know that there are high winds at the site, we also need to know from where they come, and this is where

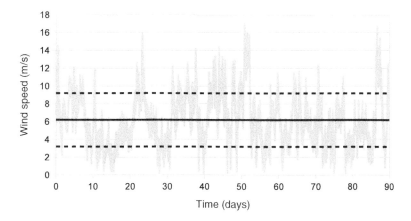

Figure 3.2 The time series of wind speed (on the y-axis) plotted versus time (x-axis). The solid straight line is the mean/average and the two dashed straight lines are plus and minus the standard deviation, respectively

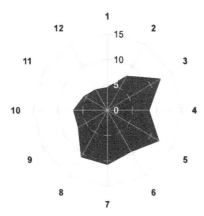

Figure 3.3 Wind rose of the example data. The wind rose shows the frequency of occurrence for each directional sector. Sector 1 is North, that is, it gives the frequency of the winds coming from North. Here the directions have been split up in 30° sectors, as an example North is from 345 to 15°

direction comes in. A simple way of plotting the directional distribution is the so-called wind rose. The wind rose for the sample data is shown in Figure 3.3. Again it is important to stress that the wind direction distribution in this plot will not be climatologically representative of the distribution at the site; it is just meant as an example. Note that in meteorology, the direction is defined as the direction *from which* the wind comes; so a westerly wind (270°) means that the wind comes from the west.

The great advantage of the wind rose is that it gives a very intuitive impression of the wind direction distribution, the main disadvantage is that it does not say anything about the wind speeds. There are various more advanced versions of the wind rose that attempt to give this information in various ways.

Another very simple way of plotting the relation between speed and direction is to plot one versus the other, that is, for each point in time, plot the point defined as $(x, y) = $ (speed, direction), an example of this, using our data set, can be seen in Figure 3.4, where you can see from which directions the high/low winds come. You can also see the wind rose (through the number of points in each direction looking horizontally), and more details about the distribution of the wind speeds, looking vertically.

Exercise 3.4   *What does the density of the points in Figure 3.4 represent?*

As you can see, a lot of information about the distribution of the data can be gleaned from this representation, but one important dimension is missing: time.

The last statistical/mathematical analysis, that can/should be done to your data, I will discuss here, is *distributions*. A distribution is a plot of the frequency of occurrence of a given wind speed, that is, how often does the wind blow at a given wind speed. In the case of our data set we find, as an example, that the wind blows 6 m/s 13% of the time. Plotting the entire distribution for our data set, we get the plot shown in Figure 3.5.

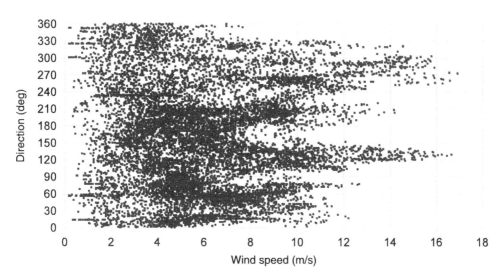

Figure 3.4   Speed (along the *x*-axis) versus direction (along the *y*-axis) for the example data. Each point represents the speed and direction pair of a given date and time

In many cases, but not all, a climatologically representative time series for a given directional sector would follow the *Weibull* distribution (see Box 7 for further details) and if you compare the plot in Figure 3.5 you can see that, even though the data is only taken over 3 months, the distribution is already very close to a Weibull distribution.

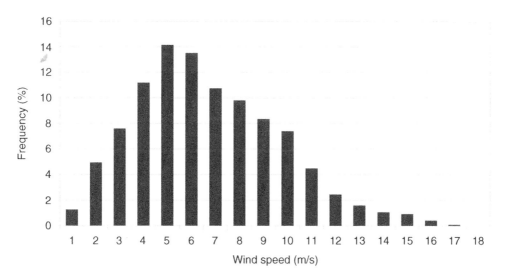

Figure 3.5   Distribution of the wind speeds of our example data set. Wind speed along the *x*-axis, frequency of occurrence (in %) along the *y*-axis

Box 7   The Weibull distribution

The Weibull distribution, named after Waloddi Weibull who described it in an article many years ago, not thinking about wind speed distributions (Weibull, 1951), can be written as follows.

$$f(u) = \frac{k}{A} \left(\frac{u}{A} \right)^{k-1} \exp \left(- \left(\frac{u}{A} \right)^{k} \right) \tag{3.2}$$

where $f(u)$ is the frequency of occurrence of a given wind speed u, k is called the shape parameter and A the scale parameter. When $k = 2$ we get the so-called Rayleigh distribution, which is also often seen in wind energy contexts.

To illustrate the meaning of k, I have plotted a few functions where A has been kept constant ($= 10$), and k varied.

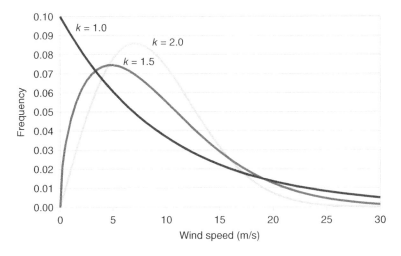

As you can see, by varying k, you can get quite different shapes of the curve, making the Weibull distribution very versatile.

A quick rule of thumb for A is that the mean of the distribution is approximately 0.9 times A.

Exercise 3.5   *Just to repeat, why is the example data series we have used not climatologically representative?*

3.3.1   Are We Good to Go?

We have now seen the many different ways that observations can and should be analysed in order to get a deeper understanding of what they show and of the data's quality. Doing these extensive analyses, which requires lots of data handling and maths, can be very difficult and time consuming, and I have seen time and time again, that people then forget to ask the very simple but most crucial question:

Are we good to go?

This means are we, after our analysis, etc., satisfied that the data we have in hand can be used for the intended purpose? The purpose can be many things, for example, finding the wind resource at a site, understanding the different stability regimes at the site, validating a model, exploring an extra parameter for a model, etc., and each of these purposes will have different sets of requirements to the data that needs to be fulfilled. So please, before you start using your data, think about that question.

3.4   Practice

In the remainder of this chapter we will turn to the discussion of the more practical aspects of measurements. This discussion will mainly cover two areas: the setting up of the masts and instruments and a tour of most of the instruments you will meet when you work with measurements related to meteorology and to wind energy in particular.

3.4.1   Measuring

You might be getting a bit tired of this caveat, but again in this part, I will only scratch the surface of the practice of measurement. I will cover enough to give you a fairly solid overall understanding, but do not consider this a 'how-to' guide to measurement in the field.

3.4.1.1   The Meteorological Mast

Most (but not all) of the instruments we use in wind energy are mounted on a mast. The main purpose of the mast is to get the instruments 'up in the air', to measure the conditions seen by the wind turbine, that is, being close to or within the rotor-swept area. A secondary purpose is to collect all the instruments in one place for practical purposes, which is why, for example, barometers (pressure sensors, see later) are also mounted on the mast even though they don't really need to be.

There are two basic types of masts: the *lattice* and the *tubular mast*. The tubular mast, which consists of a thin cylinder of typically aluminium or steel, is light and relatively easy to erect, but can generally not go as high as the lattice tower, currently 80 m seems to be about the maximum height of a standard tubular mast. The lattice tower (see Figure 3.6), which can go much higher, is also more difficult to put up and the cost is higher, too, but they are very sturdy once up, and are considered best practice for longer term wind resource measurement.

For both mast types, but in particular for the lattice tower, the flow is distorted as it flows around the structure. The distortion means that very near the tower the flow is not representative of the flow we would like to measure (which is the free stream flow), the direction as well as the speed are affected. To avoid this, the instruments used to measure speed and direction are mounted, not directly on the mast, but on a *boom*, sticking out from the mast. There are

Figure 3.6 An example of a tall lattice tower, the 125 m Risø mast. Heavily instrumented, with instruments on booms at many heights. Source: Landberg 2015

numerous suggestions and procedures describing how far away, in which direction and how thick (or rather thin) the boom should be, but the one most people refer to is the one specified by the IEC (actually in connection with the so-called power performance measurements; see IEC, 2005), such a setup is said to be *IEC compliant*. There is of course flow distortion around the booms, too; so some of the instruments are mounted on a *pole* that sits on the boom. Before this continues to infinity (!), the pole is considered so thin that the influence of it on the flow can be neglected (c.f. Figure 3.7).

Once the mast has been erected and the instruments mounted, the measurements start; this is often called the *measurement campaign*. The idea of the campaign is, as we discussed above, to measure for so long that the data we obtain is representative of what we were interested in knowing about (examples could be the long-term mean wind speed, the wind rose, etc.). This means that the mast and the instruments will be on the site for a very long time, being exposed

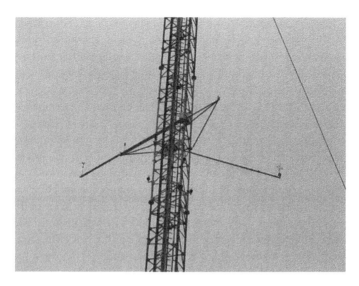

Figure 3.7 Close-up of section of a lattice tower with booms with instruments (anemometers, thermometers, sonics) mounted on poles. Source: Landberg 2015

to all kinds of winds, temperatures and precipitation. A main point of putting up a mast in wind energy is to measure the wind speed, and in general, the higher the better; this means that all masts are designed to withstand high and very high winds, so normally the winds are not a problem for the mast. Despite this, we have seen and, I am sure, will see, cases where the masts were designed incorrectly (or assembled and erected incorrectly) resulting in structural failure – it is indeed very sad to see a mast that has been blown over by strong winds!

The biggest threat to the integrity of the mast and the instruments is precipitation combined with low temperatures – some instruments are also affected just by precipitation/rain (e.g. the description of Lidars and Sodars below), but for most it is the combination that causes the problems in the form of *icing*. Icing can have from very little to a significant effect on the measurements. Before we even get to the instruments, the fact that the mast, both lattice and tubular, gets iced over can alter the flow around the structure, thereby affecting the measurements. And the ice, if thick enough, can also affect the integrity of the structure. The most important effect, however, is the icing of the instruments. We will shortly discuss the cup anemometer, and using that as an example, we can see the whole range of effects. Even the smallest amount of icing on the cups will change the aerodynamics of them, and as a consequence the readings will be incorrect (and most likely too low). This might just be a very small effect, and that can actually turn into a big problem, because there is a real danger that it will go unnoticed, biasing the numbers as a result. The more icing that gets on the cup, the stronger the effect will be (and not only on the cups, but also on the other instruments), making it easier to spot. At the furthest end of the range, the cup will be fully iced over, and be brought to a complete stop. This will of course create a very clear pattern in the data, which is good, but trying to interpolate/pad the iced data will be very difficult and in all cases introduce increased uncertainties.

The icing of the mast proper can be quite difficult to do anything about, however, the icing on the instruments like the cup anemometer, can actually be dealt with in various ways. The typical way is to use *heated* instruments. These require heating and heating requires power and in many cases this can cause a problem since getting enough power to very remote sites to heat the instruments can be quite difficult. We will shortly discuss the cup anemometer, the sonic and the wind vane, and all of these instruments can be found in versions where they are either fully or partially heated.

There are other causes that can have a similar effect (also known from the wind turbines themselves) and they include: sea/salt-spray, insects, dust, sand, dirt and so on. This means that in order to really understand the measurements, one will also need to know and have some kind of idea of what the consequences of these so-called environmental effects are.

We have now discussed the masts and in the following will discuss many of the instruments that are used for various purposes in wind energy. We will start with the most well known, but will also cover some of the lesser known. In the summary at the end of the chapter, you will find an overview of all the instruments that have been discussed.

3.4.2   Cup Anemometer

The first of the instruments we will discuss is the cup anemometer. It is called a cup anemometer because it is made up of, normally three, cups. The cup anemometer measures, as the second part of the term indicates, the wind speed more specifically the horizontal component of the wind speed. An example of a cup anemometer can be seen in Figure 3.8. As you can see, it is a very simple and sturdy instrument and it is the wind resource measurement work horse of the wind energy industry. We will see more advanced ways of measuring the wind speed below, but I think it is still safe to say that very close to 100% of wind speed measurements

Figure 3.8   A cup anemometer with three cups. Source: Landberg 2015

in connection with wind farms are carried out by cup anemometers. It is a very reliable and sturdy instrument and it requires no power to work, but of course a little to store the data electronically.

The principle behind the cup anemometer is that it transfers the movement of the air, through the rotation of the cups to a small electric generator, which in turn produces a voltage that can be translated into a measure of the wind speed.

Exercise 3.6   *This might be a simple question, but looking at Figure 3.8 which way does the cup anemometer turn, and why does it only turn one way?*

In order for the cup anemometer to measure the wind accurately it needs to be *calibrated*. Virtually all instruments need some kind of calibration to connect what they measure to some kind of standard. Some are very simple, like a scale, which just needs to be referenced to the IPK (International Prototype of the Kilogram, BIPM, 2015) in Paris (or one of its many national copies). In the case of the cup anemometer, a wind tunnel or *in-situ* calibration is normally carried out. A *wind tunnel* is a normally small tunnel (a few metres wide) where a flow of a constant, known, wind speed can be generated. A number, usually a quite large number, of different wind speeds are then generated in the tunnel to construct the calibration curve, which is a curve that relates the output of the instrument to the wind speed in the wind tunnel. The advantage of a wind tunnel is that the quality of the wind is very high, that is, steady/non-fluctuating, making it very easy to determine the wind speed. This is also a bit of a disadvantage however, since the winds we measure in the real atmosphere are rarely very steady, due to the turbulence in the flow. One way of trying to take this into account is by carrying out *in-situ* measurements. This means that the cup is put up in the real atmosphere (i.e. outside) next to an instrument with known properties, and then data is collected from the two in order to establish the relation, an *in-situ* calibration has been carried out. The big advantage of this is that the wind is 'real', that is, atmospheric as opposed to wind tunnel-generated. This is, however, also the biggest disadvantage, because as we will see shortly, a cup reacts differently to different levels of turbulence (see Chapter 6 for more on turbulence), making it much more difficult to standardise the calibration.

Looking at the calibration in the case of the cup anemometer there is a linear relation between the rotation of the cups and the wind speed, given by:

$$u = S\ell + u_0 \tag{3.3}$$

where u is the wind speed, S the rotation of the cups, u_0 the offset speed (typically around 0.1 m/s) and ℓ is called the calibration or response distance. Note that some people call u_0 the starting wind speed, but that is not correct; the starting wind speed (i.e. the wind speed where the cups start to rotate) is normally higher than the offset speed, which should be more thought of as a mathematical parameter, part of the linear fit. Returning to ℓ, you can see that it relates the rotation to the wind speed, and different cup anemometers would have different ℓ's (and u_0's).

Once the cup has been calibrated, a *calibration certificate* is issued, detailing the calibration and identifying the cup anemometer. There are various bodies that can be used for this, and a good example is the group of institutes that have formed MEASNET (MEASNET, 2015).[3] It is therefore important that the certificate follows the instrument, and that you make sure to check (or have checked) that everything is in order.

One final thing on calibration: as mentioned above, the instruments will sit on the mast for a very long time, being exposed to the winds and the weather in general, which may lead the calibration of the instrument to change, drift as it is called. To make sure that this is taken into account, the instrument needs to be re-calibrated at regular intervals. The interval depends on a lot of things, but good practice is to do it every 12 months.

We now have a simple picture of how a well-calibrated cup anemometer works, but as you can imagine, it is possible to dig one level deeper by doing more detailed studies. The one thing that needs to be highlighted here is the problem of *over-speeding*. As mentioned when discussing *in-situ* calibration, the atmosphere is turbulent; for measuring with a cup anemometer, this means that the wind speed fluctuates constantly. If the anemometer is able to respond to these changes instantly, this would not be a problem; however, this is not the case. It takes a bit of time before the cups spin up to the rate forced by the wind, so there is a lag. Unfortunately, this lag is not the same when the wind speeds up and when it speeds down. It takes the cups longer to speed down than up (due to the inertia). This means that a bias is introduced, resulting in the mean speed being slightly higher than it is in reality, and this is what is called over-speeding. The magnitude of this over-speeding is dependent on the cup anemometer, and it is possible to design a cup so as to minimise this effect.

If you want to know everything there is to know about cup anemometers, described in a very mathematical way, then you will need to consult Kristensen (1993) and I think the title says it all: 'The Cup Anemometer and Other Exciting Instruments'!

3.4.3   Wind Vane

The next instrument we shall look at is the wind vane. The wind vane measures the direction of the wind. We are still in the very simple end of instruments, and compared to the cup anemometer it is even simpler.

Traditionally, the way the wind vane measures is by the wind forcing the vane in the direction it blows, and then the position of the vane is read via the voltage that comes out of a potentiometer connected to the vane. Think of a potentiometer as the dial on a radio controlling the volume, as you turn the dial you regulate the volume. Similarly for the wind direction, the voltage is directly propositional to the wind direction. There are also more modern ways of reading the movement and position, using LEDs and phototransistors. Returning to the potentiometer types, you might know that it is not possible for a potentiometer to go all the way round (just like it is the case for the volume dial), there needs to be a gap in order for the system not to short circuit. This means that there is a certain sector, typically around 5° wide, where it is not possible to record any signal and therefore no direction; this is called the *dead band*. There are two ways of dealing with this: leaving the gap, meaning that there will be no direction data from that sector, or scaling the data up, using a factor so you end up

---

[3] As a matter of disclosure, the company I currently work for is a MEASNET member.

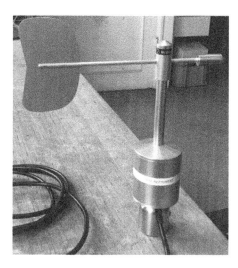

Figure 3.9   Wind vane. Vector instruments. Source: Landberg 2015

with numbers from 0 to 359 anyway.[4] If you are dealing with data at this level, it is of course important that you know what has been done to the data, including the direction data.

The resolution of the wind vane is typically a few degrees. There is very little calibration that needs to be done, but the wind vane needs to be aligned to north. This might sound simple, and it is to some extend. The problem, however, is that there are three different norths! They are the true north, map/grid north and magnetic north. True north (also called geodetic north) is the direction towards the geographic North Pole. Magnetic north is the direction a compass points towards (i.e. towards the magnetic North Pole, which is different from the geographic one). Grid/map north is the direction northwards along the grid lines of a map projection. This might sound like a very peculiar point, but in many places of the world, there is a big difference between the three directions. And imagine having aligned the wind vane to magnetic north, creating a magnetic-north wind rose, etc., and then wanting to lay-out the wind farm using a map (with north meaning the grid/map north), carefully optimising positions, wakes and such, and then realising that the two norths do not agree, and the whole wind farm has been misaligned by 20°!

We have now seen the two most important instruments in wind energy resource measurement: the cup and the vane. There are probably many, much more than half I would say, measurement campaigns where those are the only two types of instruments, and understanding how they work and the potential issues with each of the instruments cover the most important aspects of wind-resource-related measurements. However, there are many more instruments being used and with a potential of being used within wind energy. One example is that recently there has been a lot of development in the field called remote sensing. We also still need to describe some instruments that can be used to measure some of the variables we saw in the

---

[4] This does not seem quite right...

previous chapter; and finally, there are various more advanced instruments that can measure wind speed and direction. We will start with an instrument that can measure speed as well as direction, and then cover the other types after that.

3.4.4   Sonic Anemometer

The sonic anemometer can measure wind speed as well as direction. Compared to the cup, it is a much more advanced instrument, with respect to the design, the principle, and also with respect to the data processing necessary to get the values out.

Whereas the cup anemometer used the 'push' from the wind to rotate, the sonic 'listens' to the wind instead. This is done by sending out a sound pulse and then recording when that sound pulse arrives at a receiver. There are different types of sonics depending on how many dimensions of the wind velocity vector they are to measure, but they are all constructed in the same way, by having sets of transducers that alternate between transmitting and receiving the sound pulses. Calculating the time it takes for the pulse to travel along the line between the two transducers is quite simple, since the sound is carried from one to the other by the speed of sound plus the speed of the wind. Mathematically this can be written as:

$$t = \frac{L}{c + u} \tag{3.4}$$

where t is the time it takes, L the distance between the two transducers, c the speed of sound, and u the air speed along the transducer axis. The speed of sound is, as you might have wanted to point out, dependent on, amongst others, the temperature, and in Box 8 you can see how that is dealt with in a clever way, to obtain the wind speed. The transducers emit the sound waves in the ultrasonic (i.e. at frequencies higher than the human ear can hear) part of the sound spectrum.

Box 8   A bit more on the sonic

Equation 3.4 assumed that we knew the speed of sound. As mentioned above, this depends on temperature, but also pressure (and also on what kinds of contaminants there are in the air such as dust and fog). The reason the transducers alternate is that it makes the setup independent of the speed of sound, and thereby avoiding this dependence. To understand how that works, we need to look at the equation again. You can see that if the sound goes the other way, the speed is given by $c - u$, and assuming that L and c are constant in between the alternations, then the speed can be obtained from:

$$u = 0.5L(1/t_1 - 1/t_2) \tag{3.5}$$

where t_1 and t_2 are the times it takes going each way. So, no dependence on the speed of sound. Doing the same exercise for the speed of sound c, we can also get an estimate of the air temperature, via the temperature dependence of c.

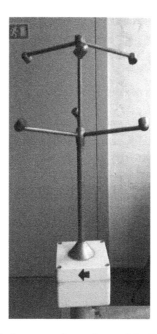

Figure 3.10   Sonic with three sets of transducers. Metek. Source: Landberg 2015

There are a number of advantages to measuring with a sonic: first, there are no moving parts, virtually eliminating wear and tear of the instrument; second, they can measure with very high temporal resolution (typically at a frequency of 20 Hz (meaning 20 recordings every second), but some types can go up to 100 Hz), which is essential if we are interested in measuring turbulence (see Chapter 6). The main drawbacks are that they require significantly more power than the cups and also that the electronics can require quite high levels of maintenance, and, as with the cup anemometer, many sonics are also prone to the calibration drifting.

Later on, we shall discuss stability (see Section 4.8) and a final advantage of the sonics is that they are able to measure the fluxes directly. A final disadvantage is that, as you can see from the photograph, there is a risk of the instrument, mainly the heads and the arms, affecting the flow, careful design and mapping of these effects in a wind tunnel, can go a long way to minimise this.

3.4.5   Hot Wire

Another way to measure the wind speed is by using a *hot wire*. This is also, in principle, a very simple instrument, in that the speed of the wind is derived from how much the winds are cooling a heated wire.

The hot wire is exactly as the term says: a hot wire, typically the wire is heated to above 300°, and kept constant at that temperature. When the wind blows harder, more current is required to keep it at the constant temperature, and by measuring the electrical signal, we get an estimate of the speed of the wind.

I guess this is a fairly unusual instrument to include here, but the reason I have included it, is that it was the first instrument for measuring wind speed that I personally used, admittedly in a very scientific context (see Box 19 if you want to know more and see an example of a hot wire in action).

3.4.6   Pitot Tube

The last of the more scientific wind-speed-measuring instruments is the Pitot tube. Named after the French engineer Henri Pitot, the Pitot tube is simply a tube with a pressure sensor at the end, the higher the wind speed the higher the pressure, subtracting the static pressure, we end up with the dynamic pressure which is proportional to the square of the wind speed ($p_d = \frac{1}{2}\rho u^2$), so recording the pressure and knowing the static pressure, we can calculate the wind speed. In wind energy we mainly know the Pitot tube from wind tunnels, where they are used to determine the reference speed, for for example, cup anemometer calibration. It is very likely that you have seen a Pitot tube near the cockpit of an aeroplane, since they are widely used in aviation (see e.g. Figure 3.11).

3.4.7   Thermometer

As mentioned above, we have still not described all the instruments we need in order to measure the fundamental variables of meteorology (velocity, temperature and pressure), so the next instrument we will describe is the *thermometer*. Well known to most, a thermometer measures the temperature of the air by using the simple physical fact that many materials expand when they are heated. Having mercury, in a thin tube, it is possible to trace the temperature by reading the height of the expanding and contracting liquid.

Figure 3.11   Two Pitot tubes on the side of an aeroplane. Source: Landberg 2015

Figure 3.12   Thermometer with radiation shield. Young multi-plate radiation shield, model 41003. Source: © R. M. Young Company. Reproduced with permission of R. M. Young Company

   I don't know if you have noticed this, but if you have a normal thermometer outside you will sometimes see very high temperatures, often unrealistically high. This is due to the fact that the thermometer does not only get heated by the air, it is also heated directly by the radiation from the Sun. For scientific and other more serious purposes this does of course not make sense, since we are only interested in the temperature of the air. The way to avoid this direct heating of the instrument by the Sun is either to put it inside a shaded well-ventilated box (called a Stevenson screen) or to put a solar-radiation protection shield directly on the thermometer, c.f. Figure 3.12.

   It is also possible to measure the *relative humidity* by using two thermometers, a normal one and one wrapped in a wet wick or sock. This might sound odd, it is, and it gets even odder, because in order to carry out the measurement, the instrument needs to be spun around (in order for the water to evaporate). This is called a *sling psychrometer* and it measures the wet bulb and dry bulb temperatures,[5] from which the relative humidity can be derived.

   It is possible, but also a bit difficult, to use two thermometers to measure the temperature *difference*, the so-called differential temperature. This is difficult, since as we saw in the previous chapter, the temperature only changes by about 1°for every 100 m, and measuring the temperature difference over a few tens of metres would mean that we are very close to the resolution and accuracy of the instruments, making accurate estimates difficult. A better, more accurate, way of doing this is to use thermocouples, which are instruments designed to measure temperature differences.

---

[5] 'bulb' refers to the bulb-shaped end of the thermometer.

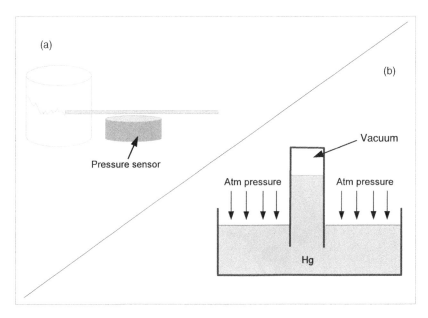

Figure 3.13   Two types of barometer principles. (a) The variation of the pressure is recorded tracing the movement of the can. (b) The pressure can be inferred from the height of the column of mercury (Hg)

3.4.8   Barometer

The last of the classic instruments is the barometer. The barometer measures the atmospheric pressure. As illustrated in Figure 3.13 there are basically two ways of doing this: by recording the changing pressure through the movement of a very pressure sensitive 'can' (Figure 3.13a) or by measuring the height of, typically, a column of mercury (Figure 3.13b).

Pressure is probably the variable that is the least sensitive to local conditions of various sorts, like heat and radiation in the case of the thermometer and local flows in the case of wind speed and direction (see Chapter 5), which is why traditional meteorology puts so much weight on this particular parameter.

It should be noted, that for the barometer as well as the thermometer, you can get 'modern' digital versions, which use advanced electronics and semiconductors, but in order to understand the principles, which is the purpose here, the classic instruments are much more appropriate.

In the following sections we shall look at some of the more advanced instruments, starting with two remote sensing devices, the Lidar and the Sodar.

3.4.9   Remote Sensing

In this section, we shall discuss two types of remote sensing devices: the Sodar and the Lidar. One is based on emitting and receiving sound waves and the other, light waves. However, before we start discussing the instruments in detail, there is an important principle that they both are based on, the Doppler shift, and we will start by understanding what that means.

3.4.9.1 The Doppler Shift

The basic idea behind the Doppler shift/effect is that we, by looking at the change of frequency of a signal, can determine the speed of what sent out or reflected the signal.

A more technical explanation is that, if you imagine a moving object emitting (sending out) sound (a much used example is an ambulance and its siren), then when the object is moving towards the receiver (a person hearing the siren), the frequency (the pitch) will increase when the object is moving towards the receiver, and decrease when moving away.

Hand-waving-wise and as a simple illustration of how this works, imagine a very simple example where two persons stand back to back, throwing balls in opposite directions, one every second. The speed of the balls is 10 m/s. Imagine further that we have two people, A and B in Figure 3.14 catching the balls, each 100 m away. It will take each ball $100/10 = 10$ s to reach the two catchers, and they will receive one every second. The details of the first few balls are show in Table 3.1.

As you can see, A receives one ball every second. Imagine now, that the two throwers start to move towards catcher A with a speed of 5 m/s. At first catcher A will catch the first ball after 10 s (i.e. the ball will arrive at $T = 10$ s), but after the first second, the two throwers have moved closer to catcher A, and the distance is now $(100 - 5 = 95$ m$)$, and the next ball will only take $(95/10 =)$ 9.5 s to reach catcher A (i.e. it will arrive at $T = 9.5 + 1 = 10.5$ s (because it was thrown at $T = 1$)); after another second the throwers will be at a distance of $(95 - 5 =)$ 90 m and it will take that ball $(90/10 =)$ 9 s to reach catcher A (i.e. it will reach catcher A at $T = 9 + 2 = 11$ s (because it was thrown at $T = 2$)). So despite the fact that the balls are only

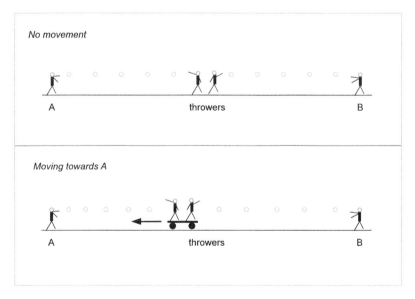

Figure 3.14 Schematic illustration of the Doppler shift with two throwers in the middle, and two catchers, A and B. Not to scale. The top panel shows the stationary case, the bottom the moving case. See text for more details

Table 3.1 Details about the throwing of the balls in the stationary case, see text for details

No move			
Time thrown	Position	Time to A	Time when at A
0	0	10	10
1	0	10	11
2	0	10	12
3	0	10	13

thrown every second, they reach catcher A every 0.5 second, because of the movement of the throwers. The details of the first few balls are shown in Table 3.2.

This is the Doppler effect/shift: if we know how often the balls are thrown, we can infer the speed of the thrower by seeing how often the balls arrive at the catcher.

Exercise 3.7 *What happens at catcher B?*

We are not really interested in balls being thrown, but rather waves (either sound or light). So imagine that the balls are the tops of the waves. This means that to catcher A, the wave length (here the distance between the tops) of the waves will seem shorter compared to what the throwers would see, which in wave language means that the frequency has changed (increased).

Finally, looking at the mathematical expression for the Doppler shift it says:

$$\Delta f = \frac{\Delta v}{c} f_0 \tag{3.6}$$

where Δf is the difference between the emitted frequency, f_0 and the received frequency, f, Δv the difference in velocity between the emitter and receiver (and equal to the velocity of the emitter, if the receiver is not moving), c is the speed of sound/light and f_0 the emitted frequency. So, since we know c and f_0, it is possible to infer the velocity of the emitter, by measuring how much the frequency has changed compared to when it was sent out.

Table 3.2 Details about the throwing of the balls in the moving case, see text for details

Moving 5 m/s towards A				
Time thrown	Position	Distance to A	Time to A	Time when at A
0	0	100	10.0	10.0
1	5	95	9.5	10.5
2	10	90	9.0	11.0
3	15	85	8.5	11.5

This was a very long journey to get to the Doppler shift, but since it is such an essential part of understanding how the Lidar and the Sodar work, I have chosen to do it in this way. We are now almost ready to start describing the two instruments, but before we do, there is one more technicality we need to understand.

Whether the instrument is based on emitting and receiving light or sound, we can only get information *along* the path of what is emitted (just as it was the case looking at the balls), this is also called the line-of-sight. And that creates a small problem: imagine that your instrument stands on the ground and sends the light/sound directly upwards. This means that the speed we can infer from the change in frequency is the *vertical* component of the wind speed, not exactly what we were looking for, which was the *horizontal* component. In order to overcome this problem, the Lidars scan in a cone, getting a projection of the horizontal wind, in each part of the cone, which can then be put together to the full horizontal wind.[6] The Sodar also uses non-vertical beams.

We now know about the Doppler shift and the problem of the non-horizontal projections, and are therefore ready to discuss the two instruments, which we shall do in the following.

3.4.9.2 Lidar

Lidar is short for LIght Detection And Ranging,[7] and the Lidar is used to measure the wind speed profile based on the Doppler shift (see section on the Doppler shift). Lidars use – eye-safe – lasers as the light source, the wave length is around 1 μm. The Lidar is ground-based, meaning that no mast is required and that a Lidar can be put up and moved in a very short time. These instruments are called remote sensing for a very good reason, since they can measure the wind speed up to 200 m above ground, which is even higher than most masts. Another more recent use of Lidars is to put them offshore on some kind of floating platform – there are of course a number of technical challenges that need to be overcome in order to do this, but the great advantage is that they replace a mast; onshore masts are expensive, offshore masts very expensive.

To detect the motion of the wind, the laser beam is reflected off of the various natural aerosols which will have the same speed and direction as the wind.

There are two types of Lidars: pulsed and continuous wave (cw). The pulsed type (see Figure 3.15) sends out pulses of light, and the height of the measurement can be inferred from the arrival time of the reflected signal. In the case of a cw Lidar (see Figure 3.16), the laser beam is focused at different heights, which are then changed in rapid succession to get a measurement of the wind speed profile.

Having an instrument standing on the ground measuring the wind speeds up to 200 m above ground level, might sound too good to be true, but it isn't. However, a question that very quickly arises is: how accurate are these measurements? I will not go into any great detail about this here, but suffice to say that currently they are very accurate, and in many cases with

---

[6] This assumes that the horizontal wind does not change as it passes the cone, an assumption that does not always hold, especially in complex terrain, but that is another story...

[7] The most famous of these -dar's is of course the one based on radio waves, the Radar.

Figure 3.15 An example of a pulsed lidar. Windcube v2. Source: © Leosphere France. Reproduced with permission of Leosphere France

Figure 3.16 An example of a continuous wave (cw) Lidar. The ZephIR 300 wind Lidar. Source: © ZephIR Lidar. Reproduced with permission of ZephIR Lidar

an accuracy comparable to that of a cup anemometer, which is used as the 'truth', meaning that a Lidar that can reproduce what a cup measures is considered accurate. This opens up a slightly more philosophical discussion: are the two measurements actually measuring the same wind speed and should they? The cup is measuring in a point, whereas the Lidar is averaging over quite a big volume of the atmosphere. How to interpret this is currently being debated, but it is still reasonable to validate a Lidar against cup anemometer measurements.

Before we leave the Lidars, there is one more topic that I would like to cover, and that is the use of Lidars to measure turbulence. Currently (2015) this is still a very rapidly developing field with many encouraging results, and a good reference is (Mann et al., 2012), but as I said, the field is developing rapidly.

3.4.9.3 Sodar

Sodar is short for SOund Detection And Ranging, and the Sodar measures the wind speed profile, just like the Lidar. Instead of being based on light, it is based on sound, actually a very characteristic chirping sound (typically only lasting 50 ms) hearable by the human ear. As the sound waves are transmitted (via a set of speakers) and go up through the atmosphere some of the energy/sound is scattered back. The backscattered energy is picked up by a set of microphones. It is only the small-scale thermally induced turbulence that is able to scatter the energy back. The return frequency is altered due to the Doppler shift (see section on the Doppler shift), and from this, the speed of the air at a given height can be calculated.

Using three beams of sound, it is possible to measure the three-dimensional wind speed vector. Most Sodars are so-called monostatic, which means that the transmitting and receiving antennas are collocated, that is in the same unit (c.f. Figure 3.17).

For many years there has been a long discussion about Lidars versus Sodars. I will not say much about this here, other than the two types of systems both have their advantages

Figure 3.17   An example of a monostatic Sodar. The AQ500 Wind Finder. Source: © AQSystem Stockholm AB. Reproduced with permission of AQSystem Stockholm AB

Figure 3.18 Illustration of the principle of the WindScanner system. Three Lidars are pointing at the same point in space. Source: © Technical University of Denmark. Reproduced with permission of Technical University of Denmark

and disadvantages. And both 'sides' are of course working on eliminating or reducing the disadvantages.

3.4.9.4 More Advanced Remote Sensing Instruments

In this section, I shall briefly mention two examples of recent developments within remote sensing, that look quite promising. This is another field under rapid development, so these two do by no means give an exhaustive description of all the new systems out there. The first is the WindScanner and the second a dual Doppler radar system by SmartWind Technologies.

The WindScanner (WindScanner, 2015) is a set of three Lidars that work together, aiming very precisely at the same point in space allowing the three-dimensional wind vector to be measured (see Figure 3.18). This is of course quite difficult, but the 'scanner'-part even takes it one step further by not only measuring at one point, but using the Lidars to scan an entire wind farm, very swiftly. In this way the entire flow field of an area can be measured. Depending on the technology the scanner can measure in the short range, that is, hundreds of metres and in the long range, up to 8 km.

A similar result is obtained by the SmartWind dual Doppler radar (SmartWind, 2015), but instead of using Lidars, it uses a radar-based system (see Figure 3.19). This extends the range significantly, up to 30 km in fact, making it possible to scan even very large areas, e.g. mapping the wakes of a large number of wind turbines, simultaneously (c.f. Figure 3.20).

We have now gone through quite a few instruments, and I am sure your head is buzzing! There is now only one instrument to go, the ceilometer, and then we will close the chapter with two high-flying alternatives to the meteorological mast.

3.4.10 Ceilometer

The ceilometer measures the height of the various cloud bases as a laser beam shines up through the atmosphere up to as high as 15 km (c.f. Figure 3.21). As the beam is reflected in

Figure 3.19 A Texas Tech University mobile Doppler Radar system mounted on a truck and deployed near a commercial wind farm. Source: © Texas Tech University. Reproduced with permission of Texas Tech University

the various layers, the receiver picks up the return signal, much like how the Lidar worked. However, of more interest to wind energy applications is that it can also measure the structure of the entire boundary layer, identifying the different parts (as described in Figure 2.6), and as we shall see in Chapter 4, knowing these makes it possible to use more advanced expressions for the vertical structure of the wind speed profile.

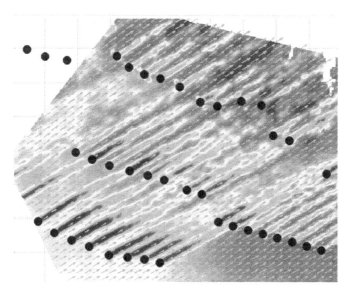

Figure 3.20 Measured wakes by the Texas Tech University dual-Doppler Radar system; this technology has since been implemented commercially by SmartWind Technologies, LLC. Source: © Texas Tech University. Reproduced with permission of Texas Tech University

Figure 3.21   An example of a ceilometer.[8]

You will be happy to know that this is the last ground-based instrument that we will describe here. As mentioned above, before we close this chapter there are two alternatives to the meteorological mast that will be described: the weather balloon and satellites.

3.4.11   Weather Balloon or the Radiosonde

One of the first weather balloons was released in 1896 in France, and since then they have been used to measure the profile of atmospheric pressure, temperature, humidity and wind speed up through the atmosphere, tracked by either radar, radio or GPS. The data is transmitted back to ground via a radio transmitter. Weather balloons are also called radiosondes (Figure 3.22).

I do not think very many people know this, but today, twice a day, every day, more than 800 of these are released into the atmosphere all over the globe, providing measurements that are used mainly for general meteorological observations and forecasting models. This is organised by the WMO and the data is shared globally. NOAA in the United States even has a page on their website describing how to return found radiosondes (NOAA, 2015b).

3.4.12   Satellite-Borne Instruments

Having left terra firma by going into the weather balloon, we will now take the final step and look at satellite-borne instruments. Satellite-borne instruments, that is instruments that are mounted on Earth-orbiting satellites, provide a lot of the data we use in wind energy (including topographic information), but wind-wise, the most well-known use is to study the winds offshore. To do this two different instruments are used: scatterometers and synthetic aperture radars (SAR).

I will very briefly describe these two instruments here: both are active sensors sensing in the microwave range (around 1 cm). Active means that they send out microwaves and then listen

---

[8] Figure source: Jk047, https://commons.wikimedia.org/wiki/File:Single_Lens_Ceilometer.JPG#/media/File:Single_Lens_Ceilometer.JPG [Public domain], from Wikimedia Commons.

Figure 3.22 An ascending radiosonde. The instruments are located in the white box at the end of the line. Source: NASA

for the signal scattered back from the surface. This is opposed to many of the other satellite data sets we see, as an example the one we saw in Figure 2.3, which is a passive system, meaning that it is just looking at what is reflected back to the Earth. Since the two instruments are actively transmitting a signal, they work day and night.

The main differences between the two are what they measure and at which resolution. The scatterometers provide ocean surface wind vectors (i.e. both speed and direction); however, the resolution is quite course with 25 km grid cells.

The SARs, on the other hand, have a much higher resolution (1 km), but they can either measure speed or direction. So we need to know either wind speed or direction to determine the other. To solve this problem, sometimes direction data from numerical models (see Chapter 8) is used. Wind streaks (basically lines on the ocean surface created by the wind) can also be used to at least get to a $\pm 180°$ idea of the direction.

How can we infer the wind speed and direction from just looking at the ocean surface from space? The answer to this question has to do with the fact that the so-called gravity-capillary waves, which are very small waves (wavelengths of 4–7 cm) formed on the surface of the water, respond instantaneously to the local wind, and the backscattered signal from these waves can be related to the wind speed (at 10 m) through an algorithm. One of the most widely used of these algorithms is the CMOD-4 algorithm (Stoffelen and Anderson, 1993).

Figure 3.23 An example of a satellite. SeaWinds scatterometer seen on NASA's QuikSCAT satellite. Source: NASA

Satellite data is mainly used for offshore wind resource estimation (see Hasager et al., 2005),[9] but another interesting application is to determine the wakes behind offshore wind farms (see Hasager, 2014).

As an example of one of these instruments, the SeaWinds instrument, which is a scatterometer, flown on the QuickSCAT satellite is shown in Figure 3.23.

3.5 Summary

In this chapter we have discussed two aspects of carrying out measurements: a more theoretical part where we looked at data, time series and time series analysis, and a very practical part where we looked at masts, the various instruments we use, including a description of the principles behind most of them.

---

[9] Disclosure: I was the manager of this group at the time they wrote the paper.

Table 3.3 A list of the instruments described in this chapter, the second column indicates what quantities they measure

Instrument	Measures
Cup anemometer	Wind speed
Wind vane	Direction
Sonic	Speed
	Direction
	Temperature
Hot wire	Speed
Pitot tube	Speed
Thermometer	Temperature
Psycrometer	Rel humidity
Barometer	Pressure
Lidar	Speed
Sodar	Speed
Ceilometer	Height
Scatterometer	Wind speed
	Direction

In the theoretical part, we started out quite philosophically by asking the fundamental question: why do we measure? The answer to that also led us in the direction of being able to estimate where, for how long, how many, and to what accuracy the measurements needed to be at, in order to fully answer the question.

We also went through some of the most fundamental aspects of time series analysis, including the mean, standard deviation, various ways of plotting data, with simple data-versus-time plots, scatter diagrams, wind roses and wind speed distributions as examples.

Looking at data it was also important to decide what was required to make the data representative, and also to distinguish between resolution, accuracy and precision.

Further to representativity, we also discussed quality, and the very important question: *are we good to go?* meaning that are we, first of all convinced that the data is of a sufficient quality (along the many dimensions we have discussed) and second, will the data help answer the question or questions we are looking to have answered. Having done a lot of detailed analysis and plotting of the data, it is sometimes difficult to remember to ask this question.

Looking at the more practical side of measuring we started out by looking at the two different types of meteorological masts, the lattice and the tubular tower. We then described a very long list of instruments shown in Table 3.3.

Exercises

3.8 *Analyse one of your own data sets, using what we have discussed in this chapter. If you do not have a data set for yourself, you could try to get hold of an external set, from for example winddata.com.*

3.9   *Find a plot of share prices, exchange rates, electricity prices or something similar and compare to our wind speed plot in Figure 3.2. Compare the 'nature' of the two: what is similar and what is different?*

3.10 *Imagine that you have two data values: 202.8 and 240.1 and you want to answer the question: are these two values equal? What kinds of questions do you need to ask to answer that question?*

3.11 *Find the mathematical expression for the Gaussian distribution.*

3.12 *Plot two Weibull distributions, both with A = 10, one for k = 2.2 and one for k = 3.0.*

3.13 *Derive Equation 3.5.*

3.14 *Why can the Windscanner measure the three-dimensional wind vector?*

4

The Wind Profile

In many ways this will be the most central chapter of the book. If you have measurements at your site, the wind profile will help you move up and down vertically; if you do not, you might have a general idea about the wind speed at the site. This wind speed will of course relate to a specific height (and Murphy's law will dictate that this will not be the hub height of your turbine); so you will need to take the wind speed from the height given to hub height, and that is where the wind profile comes in again.

The wind profile is the relation between height above ground level (agl) and the horizontal wind speed at that height. As simple as that might sound, that can be quite complicated, and this chapter will go from the most simple (and very widely used) expressions of the wind profile to some quite advanced theories and models.

One attribute of the atmosphere, that we have not discussed so far, is *stability* or more precisely atmospheric stability. This plays a very important role when more advanced wind profile models and theories are discussed, so we shall spend some time discussing this in this chapter as well.

There will also be a brief discussion about direction, even though wind direction is not as interesting as wind speed, since it, in most cases, does not change with height (in the surface layer).

4.1 A Hand-Waving Way of Deriving the Simple Log Profile

The wind profile is such a fundamental relation for wind in the boundary layer and therefore for wind energy that it is worth spending a bit of time trying to derive it. We will do this in a very hand-waving way, but there is of course a more stringent way of doing it, and I have sketched some of the idea behind the reasoning in Box 9.

In the following, we shall follow a hand-waving simple line of reasoning (and you can easily skip to Equation 4.9 if you do not think maths is fun!).

Meteorology for Wind Energy: An Introduction, First Edition. Lars Landberg.
© 2016 John Wiley & Sons, Ltd. Published 2016 by John Wiley & Sons, Ltd.

Let us start the line of reasoning with stating two facts. We know two simple things about the wind profile:

- the wind speed is zero near the surface, and
- the rate of change of the wind speed decreases with height, from being very rapid near the surface to much slower higher aloft.

This rate of change is also called the *wind shear*, and can be expressed mathematically as:

$$\frac{du}{dz} \tag{4.1}$$

So in other words, what we are saying above is that the shear decreases with height. Let us do a simple exercise:

Exercise 4.1 *Which simple physical property of the atmosphere is guaranteed to decrease with height?*

(If I could ask you to pause for a minute and think about this, please)

Looking at Chapter 2, you might think that a good answer to this question is temperature. Temperature does of course decrease with height, but it is far from uniform, and sometimes it even increases (see Figure 2.4). Pressure could be another candidate of course; but that can also vary a lot with height as well as with location. Another, a bit more strange reason, which will become clearer shortly, is that the dimensions/units of both temperature and pressure do not fit so nicely when we want to talk about wind speed and variation with height.

Okay, so what *is* a good candidate? One quantity which is guaranteed to decrease with height is the ... inverse of the height:

$$1/z \tag{4.2}$$

because when z increases, $1/z$ is guaranteed to decrease. You might feel a bit cheated by this simple answer, that was not the intention, but you will see next why it is also a very good answer.

So, let us use that as something that the shear is proportional to:

$$\frac{du}{dz} \propto 1/z \tag{4.3}$$

We can now do a bit of maths on this (isolating du on the one side and the z's on the other):

$$du \propto \frac{dz}{z} \tag{4.4}$$

If you have forgotten how to integrate, don't worry, but if we integrate both sides from z_0 (a trick!) to z, we get

$$\int_{u(z_0)}^{u(z)} du \propto \int_{z_0}^{z} \frac{dz}{z} \tag{4.5}$$

And with a bit more maths we get

$$u(z) - u(z_0) \propto \ln(z) - \ln(z_0) \tag{4.6}$$

Going back to the trick above, and defining z_0 as the height where $u = 0$, and doing a bit more maths, we get[1]

$$u(z) - 0 \propto \ln\left(\frac{z}{z_0}\right) \tag{4.7}$$

When two quantities are proportional to each other, one is multiplied by a factor to get the other, so we get

$$u(z) = A \ln\left(\frac{z}{z_0}\right) \tag{4.8}$$

In boundary-layer meteorology, convention has it that the proportionality factor, A, should be set equal to u_*/κ, where u_* is called the *friction velocity* and κ, the *von Karman constant*, equal to 0.4. We then get

$$u(z) = \frac{u_*}{\kappa} \ln\left(\frac{z}{z_0}\right) \tag{4.9}$$

which is called the *logarithmic wind profile*. Looking at real wind profiles, this equation often fits very well to observations (and it can of course be derived more theoretically, see e.g. Landberg, 1993).

Returning to our trick from before, it can now be seen that it was a good idea to have $u(z_0) = 0$ and not $u(0) = 0$, zero does not work so well with logarithms! z_0 is called the *aerodynamic roughness length* or just the roughness and we will return to it later in this chapter and also in Chapter 5. You can also find some typical values for z_0 in Table 5.1. So, we have now in a sort of pseudo-mathematical hand-waving way, derived one of the most important relations in boundary-layer meteorology!

Box 9   The physical meaning of u_*

Despite the fact that we arrived at the log profile in a very loose way, there is actually real physics behind the expression. u_* can be defined by two very fundamental quantities of the turbulent air: the vertical flux of the horizontal momentum, τ, that is, how do the horizontal winds change as we go up and down, and ρ, the density of the air, in the following way:

$$u_* = \sqrt{\tau/\rho} \tag{4.10}$$

This means that u_*, and the wind profile in turn, are closely related to the turbulence in the air (more about this in Chapter 6).

---

[1] Remember $\ln a - \ln b = \ln a/b$.

4.2   Working with the Log Profile

We have now found the relation between the horizontal wind speed and height, and in this section we shall look at this relation in a bit more detail. First, let us see what the two parameters u_* and z_0 do (we have explored the theoretical meaning of u_* in the box above). As you can see the wind speed is directly proportional to u_*, which simply means that if u_* grows so does u. z_0, the roughness, is a bit more difficult, since it is hidden in the logarithm. Going back to the 'trick' we used before, remember that z_0 is the height where the wind speed is zero, that is, when it crosses the y-axis. If we plot a logarithm in a log–linear plot (i.e. a plot where the x-axis is a normal linear axis and the y-axis is a logarithmic scale), we will get a straight line. This is useful, since the wind profile will then be a straight line in such a plot, and z_0 will be where the line crosses the y-axis. To see how the shape of the wind profile changes as z_0 changes, we will do a quick simple exercise.

Exercise 4.2   *Plot the wind profile up to 200 m for the following three cases: water ($z_0 = 0.0002$ m, $u_* = 0.31$ m/s), grass ($z_0 = 0.03$ m, $u_* = 0.43$ m/s) and a forest ($z_0 = 1$ m, $u_* = 0.58$ m/s). (we will shortly get back to why the u_*'s need to be different, basically because by doing so, we get comparable conditions for the three different roughnesses). Plot the profile first in a linear–linear plot (i.e. a normal plot) and then in a log–linear plot.*

As always, please give it a try, before you read on.

Inserting the above three sets of values in Equation 4.9 and plotting in a linear–linear coordinate system we get the plot in Figure 4.1.

As you can see, the roughness of the surface has a significant impact on the wind speed at a given height. As an example, compare the winds at 100 m for the three cases, where you can

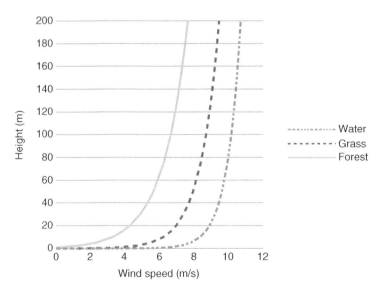

Figure 4.1   The three profiles plotted in a linear–linear plot

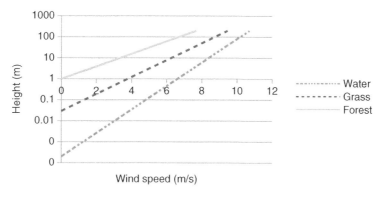

Figure 4.2 The three profiles plotted in a log–linear plot

see the wind ranges from around 7 m/s over a forest to just below 10 m/s over water. You can also see that a rough/high roughness (forest) 'holds on' to the wind much further up than the smoother roughnesses.

Plotting the same data, but with the y-axis now a logarithmic scale we get the plot in Figure 4.2. As expected we get three straight lines, and the slope is equal to u_*/κ and the height where lines cross the y-axis is the roughness, z_0.

Exercise 4.3 *As a final exercise of how to use the simple log profile, consider a mast with measurements at three heights (30 m: 5.7 m/s, 50 m: 6.2 m/s, and 100 m: 6.9 m/s). Assume that the profile follows the log profile. Find the roughness and the friction velocity. Hint: use the log-linear plot.*

Again please try yourself first.

You can do this in a few different ways, the easiest is to put the numbers into a spreadsheet, and then try different values of u_* and z_0 until you get a match. A more mathematical solution would be to select two of the sets of values and then fit the function through them. In the worked answer to Exercise 4.16 at the end of this chapter; I have done it in the mathematical way. But here, I have just varied the two parameters, and doing that we find $u_* = 0.4$ m/s, and $z_0 = 0.1$ m. A plot of the data (marked as filled circles) and the resulting profile looks like the one shown in Figure 4.3.

4.3 The Power Law

In engineering (and many other fields), the wind profile is often described by a power law:

$$u(z) = u_r \left(\frac{z}{z_r} \right)^{\alpha} \tag{4.11}$$

where u_r is a reference wind speed at height z_r, α is called the wind shear exponent (sometimes also called the Hellman exponent or the power law exponent), and $u(z)$ is the wind speed at

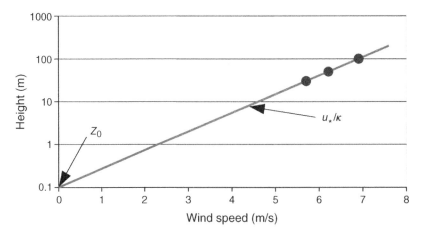

Figure 4.3 The profile plotted in a log–linear plot with the three measurement points marked as filled circles. The roughness, z_0, can be found where the line crosses the y-axis, and u_*/κ is the slope of the line

height z. As you can see this is a different way of describing the profile, we will see this later, but the log profile is based on fairly physical reasoning, whereas the power law is more mathematical/engineering in nature. The logarithmic profile and the power law look very similar when plotted, so there is of course no real difference in the two expressions, which we would expect since they are supposed to describe the same phenomenon.

A typical value of α would be 1/7 (for neutral atmospheric conditions, see Section 4.8 for what we mean by this), but it can vary from values much lower than this, and also higher. As α varies the shape of the profile changes as well. Higher α's give a profile that is flatter, that is, growing slower.

Exercise 4.4 *For a reference wind at 100 m of 10 m/s, plot the profile for the following three α's: 0.1, 1/7, 0.2.*

Please give it a try first. By inserting the three values in Equation 4.11 we get the plot shown in Figure 4.4.

As you can see, the higher the α the slower the growth. Note that had we plotted this in a log–linear plot we would also have got straight lines. Note also, that these three profiles go through the same wind speed at 100 m, as opposed to the example with the logarithmic profile, where the lines seemed to converge very high up in the atmosphere.

Exercise 4.5 *Calculate α for the measurements given in Exercise 4.2, using the 50 and 100 m values. Plot the resulting function.*

The equation for α is given by (rearranging Equation 4.11):

$$\alpha = \frac{\ln(u_1/u_2)}{\ln(z_1/z_2)} \tag{4.12}$$

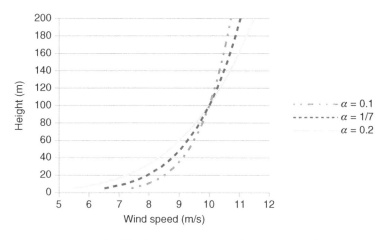

Figure 4.4   The three power law profiles plotted together

Inserting the numbers from above, we get

$$\alpha = \frac{\ln(6.2/6.9)}{\ln(50/100)} = 0.154 \qquad (4.13)$$

Exercise 4.6   *Using the α we have just derived, and the 50 m wind, predict the wind at 30 m and compare to what we had measured in Exercise 4.3.*

Since we now have the α, and two sets of reference wind speeds, it is quite simple to calculate the 30 m value (here using the 50 m values and Equation 4.11 again):

$$u(30) = 6.2 \left(\frac{30}{50}\right)^{0.154} = 5.7 \text{ m/s} \qquad (4.14)$$

Comparing this to the measurement given above (5.7 m/s), we see that there is complete agreement, which is what we would expect. Note, that if we had used the values from 100 m, then we would also have got the same value at 30 m.

4.4   Averaging Times and Other Dependencies

Before we continue with how some real profiles look, we have to take a small technical step to the side in this section, because, up until now, we have not said too much about what we actually mean when we talk about the horizontal wind speed, $u(z)$. It is of course the wind speed at a given height, z, but as you probably can imagine, and as you will hear more about in the chapter on turbulence, the wind fluctuates quite a bit all the time. So if I were to plot the *instantaneous* wind profile for, say, a given second, it would not look anything like the profiles we have looked at so far; it would be significantly more zig-zaggy due to these fluctuations. To get to the profiles we have discussed, *averaging*, as discussed in Section 3.2, is required.

Standard WMO practice is to average over 10 minutes (WMO, 2008), and doing so will often produce profiles that look more like what we would expect; however, longer averaging times are much better. But, as you probably also can imagine, if you average for too long, then the 'weather' might have changed (and with that the stability, which we will discuss shortly) changing again the way the profiles look. So, when looking at profiles, it is crucial that you understand over which period they have been averaged.

Since the 'weather' has an effect on the shape of the profile, it also has the consequence that the profile will look different from season to season (because the weather is different), meaning that a winter profile will look different from a summer profile (mainly again due to stability). Similarly a day profile might look different from a night profile.

Finally, it is rare to see a site where the roughness, z_0, is the same no matter in which direction you look, that is, it is rare that the roughness is homogeneous for the site in question. This means that the profile will look different for winds coming from different directions. Meaning again that if we average over all directions (for a heterogeneous site), we will get something which might not look like the theoretical profiles we have discussed so far.

So, to end this technical side step, in summary, make sure that you understand the averaging times, the seasons, the time of day, and the degree of non-homogeneity of your site when looking at a profile. You will, however, often find that even if you average your wind speeds over a whole year, you will still find very nice fits to the theoretical profiles we have discussed here (at least if you analyse your data by sector).

4.5   Two Famous Profiles

You will most likely meet a lot of real wind profiles in your day-to-day work, but in this section we will look at two examples of 'famous' wind profiles. The first one, one of the oldest, the Leipzig profile, is a classic and it has been used in boundary-layer meteorological research since it was published in 1932 (Mildner, 1932 and Lettau, 2011). The other, a much newer set, is the Høvsøre profiles (Peña et al., 2014). They are from 2010, and interesting because they are based on state of the art measurements taken from a tall tower.

We will take a look at these two data sets, and try to fit a logarithmic profile through them, and discuss what we find.

Exercise 4.7   *Using data from the website (larslandberg.dk/windbook, QR code in margin), plot the Leipzig profile, and fit a logarithmic profile through the points, estimating u_* and z_0.*

As always, please try to do this yourself before reading on.

Having plotted the data, and attempted to fit a log profile through the data, we get the plot in Figure 4.5. My estimate of u_* and z_0 is: $u_* = 0.7$ m/s and $z_0 = 0.15$ m (and if I use the two lowest points and does it mathematically, using the formulae in the answer to Exercise 4.16, I get $u_* = 0.75$ m/s and $z_0 = 0.23$ m).

As you can see it is possible only to fit a logarithmic profile through the first few points, since as we get higher and higher, the observations start to deviate from any kind of straight line (in a logarithmic plot). Curious why that is? See the exercises at the end of this chapter.

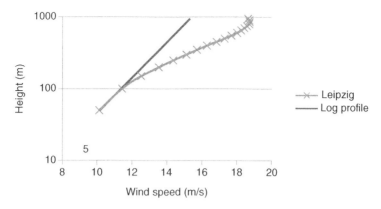

Figure 4.5 The Leipzig profile (solid line with crosses), plotted next to the standard log profile (solid line)

Exercise 4.8 *Using data from Case 6 in the article by Peña et al. (2014) about the Høvsøre data (you can find a link directly to the data from the website (larslandberg.dk/windbook, QR code in the margin)), plot the wind profile. See how the estimated values from the same paper fit the profile you have just plotted. In the paper z_0 is set to be equal to 0.015 m, and u_* 0.62 m/s (based on a direct measurement of u_*).*

Inserting the numbers from the website and the two parameters in the log profile we get the plots shown in Figure 4.6. As you can see there is a nice fit between the data and the model, for most of the heights. This means that not only does the data follow the expected straight

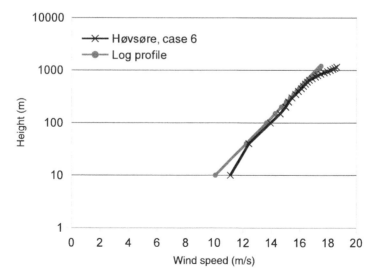

Figure 4.6 The Høvsøre profile (solid line with crosses), plotted next to the standard log profile (solid line with dots)

line, but also, through a completely independent measurement of u_*, the values are confirmed, which sort of also gives an independent validation of the theory.

The two profiles are not directly comparable as profiles go, since they are measured in very different ways: the Leipzig data was taken from 28 pilot-balloon observations with two theodolites (for triangulation) in the month of October and the Høvsøre data was measured by mast-mounted sonics up to 100 m and a Lidar thereafter (refer back to Chapter 3 for a description of these two measuring devices). The Høvsøre data was measured in early May. It is therefore very likely that the Leipzig profile has a very short averaging time (minutes at the most I would think, equal to the time it takes the balloon to pass a given height), whereas the Høvsøre profile was measured and averaged over 8 hours. The two ways of measuring are also as different almost as one can imagine, balloons versus sophisticated laser-based instruments.

As mentioned at the start of this section, you will see lots of other real-life profiles in your daily work. Remember not only to think about the profile in the framework of the theoretical profiles, but also always think about the data as measured values in the ways discussed in the chapter on measurements (Chapter 3).

4.6 Zero-Plane Displacement

As you will see in Chapter 5, the flow – to a first approximation – follows the terrain. That means that the wind profile will also follow the terrain; however, if the flow encounters a porous medium (i.e. something which is not solid – think of a forest) the flow will also need to 'go over' the medium in some way. And since the flow goes over, so will the profile (which is just another way of representing the flow). The issue now is that the height above ground level is still defined from the surface up and not from the – often ill-defined – top of the porous medium/forest. In order to take this phenomena into account, the *zero-plan displacement*[2] is introduced. Instead of writing the profile in the usual way, it is written as

$$u(z) = \frac{u_*}{\kappa} \ln \left(\frac{z - d}{z_0} \right) \tag{4.15}$$

where d is the zero-plane displacement. As you can see in Figure 4.7, the effect of d is to lift the profile, which is otherwise unchanged. Note that, if plotted in the logarithmic–linear plot the line will cross the y-axis at $z_0 + d$.

Exercise 4.9   *Why will the line cross at $z_0 + d$?*

4.6.1 Dealing with Forests

Using again the forest as an example of a porous medium, a very appropriate question to ask is: what happens at and near the forest edge? This is currently (2015) under quite a bit of investigation, involving modelling and measurements with cups and sonics as well as Lidars

---

[2] Some use the term zero-plan displacement height or length and some just zero-plan displacement.

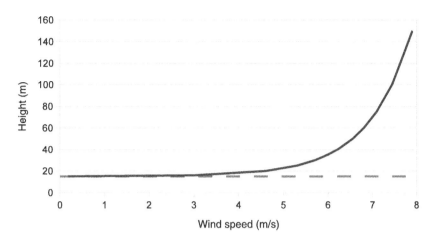

Figure 4.7 A logarithmic wind profile (solid line) in a situation where the flow is over a porous medium (e.g. a forest). The displacement height is 15 m, illustrated by the dashed line. $u_* = 0.4$ m/s and $z_0 = 1.0$ m

(see e.g. Dellwik et al., 2013, for a good overview). A simple approach, however, is the following (c.f. Figure 4.8): divide the area into three parts: far away from the forest, at the forest edge, and over the forest. It is then possible to assume that:

- Far away: no displacement height, roughness that of the surface (z_{01} in the figure)
- Over the forest: displacement height and forest roughness, d and z_{02} in the figure, respectively.
- At the edge: tapering off from the displacement height of the forest to zero, and roughness equal to that of the forest (z_{02}).

The way the displacement height tapers off is also not entirely clear, but a simple rule is to assume that it is a straight line, with a slope of 1:100 (see e.g. Corbett and Landberg, 2012).

A final question we need to answer is: what is the roughness of a given forest and what is the displacement height of this same forest? Again, this first depends on the density and the species of the trees, but, second, there is no universal agreement about this; the roughness can vary from 0.4 (which is probably quite low) to over 1 m. The displacement height is normally said to be between two-thirds to one times the characteristic height of the forest (again see e.g. Corbett and Landberg, 2012).

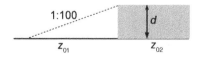

Figure 4.8 Dealing with the flow over, near and far away from a porous medium (the shaded area, think of a forest). z_{01} and z_{02} are the roughnesses of the far-away area and the forest, respectively, d the displacement height, and the line marked 1:100 indicates the slope of the tapering off-line (not to scale)

4.7 Internal Boundary Layers

Then next addition to our simple view of the vertical wind profile is the introduction of internal boundary layers (IBL). These layers develop when the roughness changes – actually every single time the roughness changes even the slightest, but in most cases the changes are not dramatic. However, imagine that you have a very simple case of a clear and significant roughness change, for example, a coastline, where the roughness would change by several magnitudes, and imagine further that the wind blows from the sea to land. When the flow is over the sea, the profile is given by our logarithmic wind profile, with z_0 equal to the roughness of the sea. When the flow is well over land, again the profile is given by the logarithmic profile, but with z_0 equal to the roughness of the land surface. The question now is: what happens right after the roughness changes? This is where the IBL starts to develop. In hand-waving terms what happens is that the flow gradually gets into equilibrium with the new roughness; at first it is only the flow nearest the surface which is affected by the new (land) roughness, but as we get further and further inland this level gets thicker and thicker, right until it reaches the height of the boundary layer after which we have reached equilibrium with the over-land roughness. The top of this developing layer is what defines the IBL. A very simple rule of thumb says that, if you are not right at the coastline, the height grows as 1:100 (Raynor et al., 1979), that is, if we go 100 m inland, the height of the layer has increased by 1 m – note, that this is quite a slow growth.

Looking at the resulting profile, this can be formulated more mathematically as (based on Troen and Petersen, 1989):

$$u(z) = \begin{cases} \frac{u_{*1}}{\kappa} \ln\left(\frac{z}{z_{01}}\right) & \text{for } z \geq c_1 h \\ u'' + (u' - u'') \frac{\ln(z/c_2 h)}{\ln(c_1/c_2)} & \text{for } c_2 h \leq z \leq c_1 h \\ \frac{u_{*2}}{\kappa} \ln\left(\frac{z}{z_{02}}\right) & \text{for } z \leq c_2 h \end{cases} \qquad (4.16)$$

where $u' = (u_{*1}/\kappa) \ln(c_1 h/z_{01})$, with the $_1$'s indicating values for the upstream (water in the description above) wind profile, and $u'' = (u_{*2}/\kappa) \ln(c_2 h/z_{02})$ with the $_2$'s indicating values for the downstream (land) wind profile. The two constants are (according to Sempreviva et al., 1990), $c_1 = 0.3$ and $c_2 = 0.09$. h is the height of the IBL (we will return to this shortly). This is quite a complicated set of equations, but if you look at the first part, you will recognise the upstream profile (the one with the $_1$'s (water)) and the last part is the downstream profile (the one with the $_2$'s (land)). It is only the middle expression which is a bit complicated, but that is the part where the one profile gradually and smoothly changes to the other.

Let us try to make sense of this through an exercise.

Exercise 4.10 *Plot the wind profile in an IBL caused by a coastline. The wind is blowing onshore. Use the following values: water: $u_{*1} = 0.3, z_{01} = 2 \cdot 10^{-4}$; land: $u_{*2} = 0.525, z_{02} = 0.2, h = 200\ m$.*

(Please try yourself first) Inserting the values in Equation 4.16 and plotting, we get the profile shown in Figure 4.9. You can very clearly see the two kinks: the first one (at 18 m, $= c_2 h$)

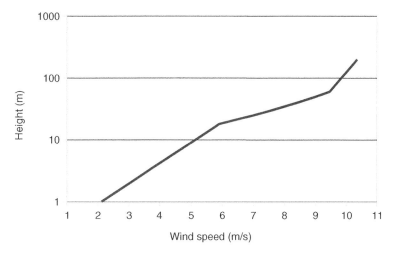

Figure 4.9   The wind speed profile resulting from Exercise 4.10

marks the end of the layer in equilibrium with the new surface (land), and the beginning of the transition zone, the second one (at 60 m, $= c_1 h$) marks the end of the transition zone and the beginning of the layer still in equilibrium with the upstream surface (water).

Returning now to the height of IBL, again (as you by now have experienced quite a few times) there is no universal agreement of what the model for this should be, and also, for the most common model, there is no agreement about what the constants in the model should be. The model which I will describe here is the one by Miyake (1965), and it states, the height, h, at distance, x, can be found from:

$$\frac{h}{z_0'}\left[\ln\left(\frac{h}{z_0'}\right) - 1\right] = C\frac{x}{z_0'} \qquad (4.17)$$

where $z_0' = \max(z_{01}, z_{02})$, and $C = 0.9$. An example is given in Figure 4.10.

Exercise 4.11   *Calculate the height of the IBL 1000 m inland, assume that the land roughness is 0.1 m*

I know this looks a bit complicated, so feel free to skip this one. As you can see from Equation 4.17 no analytical solution exists, so what I have done is used a spreadsheet to numerically solve the equation for $x = 1000$ (and $z_0' = 0.1$) m, doing that I get

$$h = 65.6 \text{ m}$$

which means that if we multiply by $c_1 = 0.3$, we get $h_1 = 19.7$ m, which is not too far from the 1:100 rule of thumb (which would have given 10 m (= 1000/100)).

Since we shall soon discuss stability, I also need to mention that the development of the IBL in more advanced models also depends on stability. In hand-waving terms, the more stable

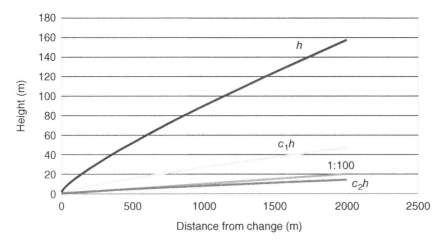

Figure 4.10 The development of an internal boundary layer (IBL) resulting from a change of roughness (at 0 m). The top line, marked 'h', is the height of the IBL, and, referring to Equation 4.16, the next line from the top is the c_1h-line, next down, marked '1:100' is the 1:100 rule of thumb. The bottom line is the c_2h-line. The x-axis is the distance from the roughness change. The upstream roughness is water, and the downstream roughness is set to 0.6 m

the atmosphere, the slower the growth (since the mixing is damped in a stable atmosphere), and vice versa for an unstable atmosphere. The profile we have looked at here is of course for neutral conditions.

4.8   Stability

A very important concept, that we have not discussed this far is *atmospheric stability*, often just called stability. The various thermodynamic equations and relations around this subject are quite complicated, but I will try to give as simple an explanation as possible – one thing I have had to do, though, is to assume that the atmosphere is dry, that is, there is no water vapour present. This is of course not correct, but introducing water vapour significantly complicates things, and by just studying the dry atmosphere, we are still able to discuss all the important concepts.

I shall come with a formal definition of stability a bit later, but basically there are three states the atmosphere can be in, and this corresponds to the following types of stability:

- Stable
- Neutral
- Unstable

Stability is closely related to temperature or rather changes of temperature with height and to understand how the temperature changes with height, imagine that there is a basic/natural way this can happen given the physical properties of our planet, our sun and our atmosphere.

The variation of the temperature with height is called the lapse rate, $\Gamma = -dT/dz$, and the natural one, mentioned above, is called the *dry adiabatic lapse rate*, Γ_d, and is given by:

$$-dT/dz = \Gamma_d \equiv c_p/g \tag{4.18}$$

T is the temperature as measured by for example, a thermometer, z the height, c_p is called the specific heat (at constant pressure)[3] and g the gravitational constant. As you can see, these two quantities are indeed quite fundamental. Note that this means that the temperature always decreases with height in this simple formulation, but as we have already seen (in Chapter 2) this is only the case in the first part of the atmosphere. Inserting the values for c_p and g, we can see that Γ_d is equal to

$$\Gamma_d = \frac{c_p}{g} = \frac{1004 \text{ J/kg/deg}}{9.81 \text{ m/s}^2} = 9.8°C/km \tag{4.19}$$

meaning that if we go 1 km upwards, the temperature will decrease by approximately 10°C, or as many hill walkers say, 1°C per 100 m. The rate of change of temperature with height of the real atmosphere, that is, the real Γ, is more like 6–7°C/km.

At this point we need to introduce a new type of temperature called the *potential temperature*, θ:

$$\theta = T \left(\frac{p_0}{p}\right)^{R/c_p} \tag{4.20}$$

as before T is the 'real' temperature at a given height, p_0 the pressure at the surface, p the pressure at the same height as T, R the gas constant (= 287 J/deg/kg for dry atmospheric air), and c_p the specific heat. This equation is also known as Poisson's equation.

This is quite a convoluted quantity, but basically it is a quantity that does not change if you move an air parcel up or down, without adding or subtracting heat to the air parcel (this is called an adiabatic process). Furthermore, it has a very useful property, in that, if the atmosphere is neutrally stratified, the potential temperature is constant with height, that is,

$$d\theta/dz = 0 \tag{4.21}$$

Furthermore, if the change with height of the potential temperature is positive

$$d\theta/dz > 0 \tag{4.22}$$

then the atmosphere is stable and if

$$d\theta/dz < 0 \tag{4.23}$$

the atmosphere is unstable. This is the most general definition of stability, and it is important to note that the atmosphere can get to this state in many different ways. In the following, however, we shall focus on the surface layer.

---

[3] Not to be confused with the wind-turbine-related C_p in the chapter on wakes.

4.8.1   Stability in the Surface Layer

Near the surface, that is, in the surface layer, the direct cause of the changes of temperature with height and thereby the resulting changes in atmospheric stability is the radiation from the Sun. In very simple terms it works as follows: the radiation from the Sun hits the Earth's atmosphere and shines right through it (this is most likely not what you have expected, but the atmosphere is transparent to the wavelengths of the light from the Sun), the radiation is then absorbed by the surface of the Earth, which in turn is heated, the (infrared) radiation that results from this heating, shines on the atmosphere from below, which is more or less opaque to this, that is, it absorbs all the infrared radiation, which in turn means that the atmosphere is heated. So it is from the ground (and not directly from the Sun) that the atmosphere is heated (see Figure 4.11 for an illustration). At night it works in the opposite way. Just to reiterate, since it is a very common mistake to assume the Sun heats the atmosphere directly, the Sun shines through the atmosphere, heats the surface, which in turn heats the atmosphere.

Box 10   Relation between T and θ

As mentioned in the beginning of this chapter, the underlying equations that form the basis of atmospheric thermodynamics are not so simple, but in this box I shall just explore the relation between T and θ a bit more, as always with these boxes, it is safe to skip them, but you will gain further knowledge if you persevere!

The relation between the change with height of T and θ is given by the equation

$$\frac{1}{\theta}\frac{d\theta}{dz} = \frac{1}{T}\left(\Gamma_d - \Gamma\right) \tag{4.24}$$

Again quite a complicated relationship, but try to see what happens when we have a neutral atmosphere, that is, one where $d\theta/dz = 0$, we then get

$$\frac{1}{\theta}\cdot 0 = \frac{1}{T}\left(\Gamma_d - \Gamma\right) \tag{4.25}$$

which after a little bit of manipulation gives

$$\Gamma = \Gamma_d \tag{4.26}$$

that is, when the atmosphere is neutral, the change in temperature with height is equal to the adiabatic lapse rate, exactly as we would have expected, and as defined above.

The equation can also be used to derive the other relations between Γ and Γ_d mentioned in the text, given $d\theta/dz$.

Exercise 4.12   *Draw the same simplified picture as in Figure 4.11 but for the situation at night.*

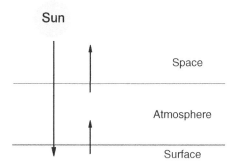

Figure 4.11 Simplified illustration of the radiation balance between space, the atmosphere and the surface of the Earth. The arrows indicate the direction of the radiation

As mentioned above, this is of course a very simplified picture of how the Earth's radiation balance works and a more detailed picture can be found in Figure 4.12. Looking at the figure, you can still find our simplified model in there of course, but just with a few more details.

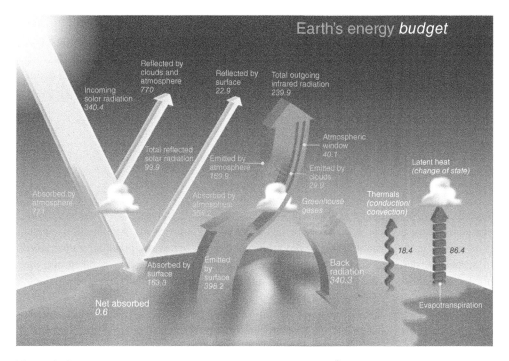

Figure 4.12 More accurate energy balance. All fluxes are in W/m² and are average values based on 10 years of data. Source: NASA's ERBE (Earth Radiation Budget Experiment) program. Reproduced with permission of NASA

Box 11   Albedo

In planetary science there is a quantity that describes the radiational characteristics of a planet and it is called the *albedo* and it is defined as follows:

The proportion of the incident light or radiation that is reflected by a surface (typically of a planet).

In simpler terms, it is how much light a planet or a surface reflects. Albedo has actually nothing as such to do with stability, but it has to do with the radiation budget described in Figure 4.12, and you will hear the term when scientists are discussing climate and climate change. We not only talk about the albedo of a planet, but also of a surface: ice would have a very high albedo (around 0.9), and asphalt a very low one (around 0.05). If you solve Exercise 4.28, then you will also know what the overall albedo of the Earth is.

An important part of the relation between the atmosphere, the incoming radiation and its temperature is the *greenhouse effect*. As we have seen, the atmosphere traps the radiation, before it radiates it back to space. This trapping results in that the atmospheric temperature is increased compared to what it would have been, had there been no atmosphere; this is called the greenhouse effect. This effect has raised the temperature of the atmosphere by more than $30°C$.[4]

Returning now to the surface, the amount of heat that comes out/goes into the surface (per unit area) is called the *heat flux*, H. Normally the heat flux is considered positive when the heat comes from the surface of the Earth and into the atmosphere. We are now able to define stability (near the surface) in a new way.

When

- $H > 0$ the atmosphere is unstable
- $H = 0$ the atmosphere is neutral
- $H < 0$ the atmosphere is stable

The connection back to the potential temperature, θ, is that the natural state we talked about is the state where no heat is added to or taken out of the atmosphere. And similarly when heat is added (in the unstable case) to the atmosphere from below the potential temperature will decrease faster with height because of the added heat, and when heat is taken out, the typical night situation, the potential temperature increases with height giving us a stable atmosphere.

Before we leave the discussion about stability, it should be noted that when we do simple modelling we almost always assume that the atmosphere is neutral, because this simplifies

---

[4] This is actually not how a greenhouse works, since it is more a mechanical trapping of the hot air that increases the temperature in there, but that is another story.

things significantly. However, as you might have realised, it is quite rare that the atmosphere is actually neutral. During the day the Sun shines at the surface, making the air warmer and thereby rising (called convection) and the atmosphere unstable and at night the opposite takes place, so in fact it is only during the brief intervals at and around sunrise and sunset that we find true neutral conditions.

The simplified models are, however, 'saved' by the fact that for slightly unstable conditions (a state which the atmosphere is in much more often) and at higher wind speeds in general (which is what we are interested in in wind energy) the atmosphere still looks neutral and as a consequence the neutral models actually work quite well.

4.9   Monin–Obukhov Theory

There is a theoretical framework that lies behind most of stability dependent boundary-layer meteorology and it is called the Monin–Obukhov similarity theory (MOST), named after the two Russian scientists Monin and Obukhov.[5] The theory was developed in 1954 (Monin and Obukhov, 1954) and was concerned specifically with how turbulence behaved in the atmospheric surface layer and in the usual Russian scientific tradition it was heavy on maths, but produced elegant results.

I will not go into any deep detail about the theory, but one of the basic assumptions is that, in the surface layer:

The vertical fluxes of momentum (and heat) are constant.

Bio 4: Alexander Mikhailovich Obukhov

1918–1989

Russia

Russian physicist and applied mathematician known for his contributions to statistical theory of turbulence and atmospheric physics. He was one of the founders of modern boundary-layer meteorology.

You will have noted that this is so ingrained in the way we think about the boundary layer that I have already called the surface layer the constant flux layer. Based on this and other assumptions, Monin and Obukhov found that there is a very important scaling parameter that describes the turbulent flow in the boundary layer called the Monin–Obukhov length (it was actually proposed by Obukhov in 1946 (Obukhov, 1946 and Obukov, 1971 (in English)) but

---

[5] This is quite a technical section, you can skip it if you so wish.

Bio 5: Andrei Sergeevich Monin[6]
1921–2007
Russia
Russian physicist, applied mathematician and oceanographer. Monin was known for his contributions to statistical theory of turbulence and atmospheric physics.

since it is so essential to MOST it is often called the Monin–Obukhov length), L, which says something about the buoyancy of turbulent flows, and it is defined as:

$$L = -\frac{u_*^3 \overline{\theta_v}}{\kappa g \left(\overline{w'\theta_v'}\right)_s} \tag{4.27}$$

As you can see, this is a very elaborate quantity, but you will recognise u_*, g and κ; θ_v is called the virtual potential temperature, and w is the vertical velocity, $_s$ means that the quantities are taken at the surface. It is hard to see, but in hand-waving terms, L (or the absolute value of) is the height where turbulence is generated more by buoyancy (the parts containing θ_v and g) than by wind shear (u_*).

Exercise 4.13   *What is the unit of L?*

$\left(\overline{w'\theta_v'}\right)_s$ is what determines the sign of L, since it describes how the vertical velocity w varies with temperature (actually the variations of the two quantities, indicated by the $'$s), so during the day when the air rises the quantity will be positive (and L thereby negative), and at night the air sinks and the quantity will be negative (and L positive). At sunset and sunrise, the quantity is zero, making L infinite. To bring this back to our stability view, we have

- $L < 0$ the atmosphere is unstable
- $L = \infty$ the atmosphere is neutral
- $L > 0$ the atmosphere is stable

due to the 'inconvenience' of having an infinite quantity, z/L is often used instead of L, making neutral conditions equal to zero.

---

[6] Figure source: P.P. Shirshov Institute of Oceanography, Russia http://www.ocean.ru/eng/images/stories/publications/monin_foto.jpg. Reproduced with permission of Shirshov Institute of Oceanography, Russia.

Using L it is possible to formulate *stability-dependent* versions of the logarithmic wind profile:

$$u(z) = \frac{u_*}{\kappa} \left[\ln\left(\frac{z}{z_0}\right) - \psi\left(\frac{z}{L}\right) \right] \tag{4.28}$$

where we recognise all the variables, and can see that we are now using z/L instead of L directly. The ψ function is not entirely well known (I have listed a few versions in my PhD thesis, Landberg, 1993), but a set that is fairly widely used is the set called the Businger–Dyer profiles (Dyer, 1974, Garratt 1992, Businger, 1988):

$$\psi = \begin{cases} -5\frac{z}{L} & z/L > 0 \text{ (stable)} \\[2mm] 0 & z/L = 0 \text{ (neutral)} \\[2mm] 2\ln\left(\frac{1+x}{2}\right) + \ln\left(\frac{1+x^2}{2}\right) - 2\tan^{-1}(x) + \frac{\pi}{2}, & \\[2mm] x = \left(1 - 16\frac{z}{L}\right)^{1/4} & z/L < 0 \text{ (unstable)} \end{cases} \tag{4.29}$$

the function proposed for the unstable cases is quite complicated as you can see, but the important point here is that we now have a framework that describes the way the wind varies with height for all atmospheric stabilities.

Box 12 Stability and smokestacks

No text about stability and meteorology without the classical pictures of smoke coming out of a smokestack under various atmospheric stability conditions.

These illustrations are very handy, since they are an easy way of estimating what the stability of the atmosphere is.

Each of the cases has also been given a name in air pollution meteorology, as indicated below.

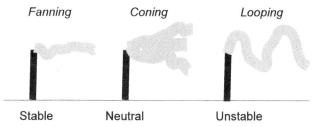

As you can see, in the stable case, there is very little up/downward movement, in the unstable a lot, and the neutral, is just letting the smoke flow.

4.9.1 Summary of the Different Stability Parameters

To conclude our discussion about stability, where we have introduced quite a few ways of describing it, an overview can be found in Table 4.1.

Table 4.1 Overview of the different ways to describe atmospheric stability. First the name of the quantity is given, then the mathematical expression and finally the sign for each the three cases

Name	Quantity	Stable	Neutral	Unstable
Lapse rate	Γ	$< \Gamma_d$	$= \Gamma_d$	$> \Gamma_d$
Potential temperature	$d\theta/dz$	> 0	$= 0$	< 0
Heat flux	H	< 0	$= 0$	> 0
Monin–Obukhov	L	> 0	$= \infty$	< 0
	z/L	> 0	$= 0$	< 0
Smokestack (Box 12)		Fanning	Coning	Looping

4.10 Deviations with Height

As the wind turbines have grown larger and larger, it has become obvious that sometimes – especially under stable conditions – most, if not all, of the rotor disc can be outside the surface layer. It is also a fact that MOST (c.f. Section 4.9) is only valid in the surface layer, since the definition of it being well mixed (and therefore constant flux – of momentum, etc.) is what is used as the basis for formulating the theory. This means that we cannot use the classic logarithmic wind profile as a description of the vertical variation of the wind as we leave the surface layer. Instead, some slightly more advanced theories and expressions are needed.

The theory I am going to describe here in only a bit of detail is that of Gryning et al. (2007) and it requires two further parameters for it to work: the height of the boundary layer (the PBL in Figure 2.6) and a quantity that relates to the flow in the middle of the PBL.

Before we go any further, we shall take a step back up into the helicopter and look a bit on what the various parameters that the vertical structure of the wind in the atmosphere is determined by.

In the simple formulation there were only two: z_0, the roughness length and z, the height. In general terms and looking at it from the view point of the entire atmosphere, flow very near the surface should only depend on the characteristics of the surface, that is, the roughness and how far away from the surface you are, z. Including stability in our formulation introduces another 'length', the Obukhov length, which is not really a length, but translates stability to a length scale.

As we get further and further up into the atmosphere, it makes sense to leave just looking at the surface and its characteristics, but also to look at the entire boundary layer. An obvious parameter is then the height of the boundary layer. This is how, hand-waving-wise, we end up with a formulation that includes the three parameters, that are used in the Gryning et al. model.

I am not going to give all the details here, but the neutral version of the formulation is given below. This area is also under quite a bit of development, but currently this is one of the best bets

$$u(z) = \frac{u_*}{\kappa}\left[\ln\left(\frac{z}{z_0}\right) + \frac{z}{L_{\text{MBL,N}}} - \frac{z}{z_i}\left(\frac{z}{2L_{\text{MBL,N}}}\right)\right] \qquad (4.30)$$

where u, u_*, κ, z are the usual variables, and z_i the height of the boundary layer (c.f. Figure 2.6). As you can see there is one further parameter, $L_{MBL,N}$, and that is describing a characteristic length scale of the middle of the PBL.

As you can see, we recognise quite a few of the terms in the equation, since it is basically an extension of the traditional logarithmic profile. The first term is our well-known log profile, and the second and third terms, are used to scale the height z with L and z_i. Higher values of either of these two terms (where infinity is the extreme case) will make the contribution of the terms smaller (and $= 0$ in the case of infinity), which is going back to the classic case, by making the atmosphere so high, that it is again only the surface characteristics that are important.

4.11   Connection with Geostrophic Drag Law

Going back to Section 2.3.3 where we introduced the geostrophic drag law:

$$G = \frac{u_*}{\kappa} \sqrt{\left[\ln\left(\frac{u_*}{fz_0}\right) - A\right]^2 + B^2} \tag{4.31}$$

you will now recognise u_*, the friction velocity. Remember that G is the wind aloft and that u_* has to do with the flow at the surface. This means that if we solve the geostrophic drag law with respect to u_*, we have found a connection between the wind aloft and that at the surface. It is unfortunately not possible to solve Equation 4.31 analytically, but it is quite easy to do so numerically (or even just by trial and error in a spreadsheet). Let us give it a try:

Exercise 4.14   *Given a geostrophic wind of 15 m/s, and a roughness, z_0, of 0.05 m. Find the wind speed at 50 m height at 55°N. (Remember A = 1.8 and B = 4.5)*

As a more general example, I have plotted the same situation as in the exercise but just for a range of geostrophic wind speeds in Figure 4.13 (with z_0, z and latitude constant). As you can see, the answer to the question is $u(50m) = 9.7$ m/s.

Exercise 4.15   *Going back to Exercise 4.2 explain why the u_*'s needed to be different.*

This exercise touches a quite important point, so I will answer it here, but as always, give it a try yourself first, please. The reason that the u_*'s are different is that that is what is required for the geostrophic wind, G, to be *constant*. This gives a more realistic picture of the effect of varying the roughness at a given location under similar conditions (i.e. the same 'weather'), which was the intention of the exercise.

4.12   Effect of Orography, Obstacles and Thermal Flows on the Profile

As you can imagine, and as you will realise after having read Chapter 5, there are many cases where the flow that arrives at your mast does not come from simple, uniform, flat terrain. Any

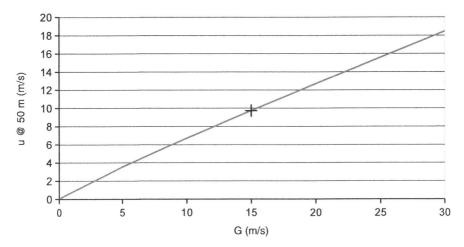

Figure 4.13 The relation between the geostrophic wind speed, G, and the wind speed, u, at 50 m agl at 55°N and with a roughness of 0.05 m for a range of geostrophic wind speeds. The solution to Exercise 4.14 is marked with the +. As you can see the relation is very close to being linear

deviation from this will have an effect on the profile, as illustrated in Figure 4.14. The list of possible situations where the simplified assumptions do not hold is very long. It includes: speed up/down effects over hills and valleys, changes in the roughness, displacement (as we have just seen), changes to the flow due to obstacles, stability (as described above) and many types of thermally driven flows.

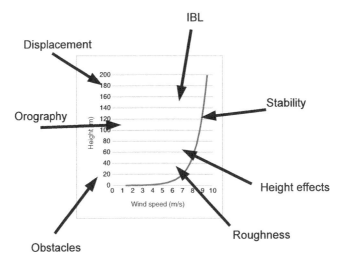

Figure 4.14 Some of the effects that change and complicate the simple logarithmic profile formulation. The effects mentioned in this chapter are part of the list, but a detailed description of the effects caused by local flows can be found in the next chapter

The consequence of this is that you need to understand the flow at your site in quite some detail, before you start looking at and analysing the wind profile. As mentioned above, all these effects will be described in Chapter 5, and in many cases, if you do some kind of modelling the model will take the particular fact into account to be accurate.

A final complication, and a fact you need to be aware of is that, as I have mentioned above, this most likely will vary from sector to sector, that is your reasoning needs to the sector/direction dependent.

4.13 Direction Profile

This is going to be a very brief section, since almost by definition (via the constant flux of momentum assumption) the direction does not change with height in the surface layer. So,

$$\theta^d(z) = \theta_0^d \tag{4.32}$$

where $\theta^d(z)$ is the direction of the wind vector at height z, and θ_0^d is the direction at the surface.

As we saw in Section 2.3.3, once we leave the surface layer, the wind direction will very much change, following (at least theoretically) the Ekman spiral.

4.14 Summary

In this chapter, we have looked at how the wind speed (and the direction only to a limited extend) varies with height. We started out with one of the most fundamental equations in boundary-layer meteorology, the logarithmic wind profile. We then added complications to get a deeper understanding of this vertical variation of the wind speed. First, we introduced the displacement (the most typical example of this was for flow over a forest). We then looked at internal boundary layers caused by the change in roughness of the surface. After that we introduced stability, which is an essential ingredient, not only in the MOST, but also for many other topics in this book. We looked at how the vertical profile deviated from the log profile as we got very high up into the atmosphere (basically outside the surface layer). And finally we saw, by knowing the geostrophic wind, how we could find the wind at the surface and vice versa.

The wind shear exponential law was also presented; this is another way of describing the same shape as the basic logarithmic profile.

As discussed in Section 4.12, it is important to realise that the profile is of course the sum total of all the various flow patterns that meet at the location where we measure. Surprisingly often one can get a very good fit to the traditional log profile, but – on the other hand – also do not get surprised if you find significant deviations. Use what you have learnt in this chapter and what you will learn in the chapter on modelling (Chapter 8) to fully understand what to expect and what is going on, whilst of course also not forgetting what you have learnt about measurements in Chapter 3.

Exercises

4.16 *Consider a mast with measurements at three heights (30 m: 6.26 m/s, 75 m: 7.41 m/s, and 120 m: 8.00 m/s). Assume that the profile follows the log profile. Find the roughness and the friction velocity.*

4.17 *Calculate α for the measurements given in Exercise 4.16, using the 75 and 120 m values.*

4.18 *As discussed in the text, try to calculate the wind at 30 m, using the α from Exercise 4.17, using the 120 m wind.*

4.19 *Why do you think the Leipzig profile (discussed in Section 4.5) starts to deviate from the logarithmic profile after 100 m?*

4.20 *Which geostrophic wind did I use in Exercise 4.2?*

4.21 *Since the atmosphere traps the heat (radiated from the surface), why does the atmosphere not get hotter and hotter and hotter yet?*

4.22 *You have measured the wind profile in a forest, and get the following:*

> *50:   4.53*
> *80:   5.76*
> *100:   5.98*

what is z_0, u_* and d?

4.23 *Plot the wind profile in an IBL caused by a coastline, the wind is blowing onshore. Use the following values: water: $u_{*1} = 0.2$, $z_{01} = 0.0002$; land: $u_{*2} = 0.35$, $z_{02} = 0.2$, h = 280 m.*

4.24 *Calculate the height of the IBL 2500 m inland. Assume that the land roughness is 0.05 m.*

4.25 *You have received two measurements of the potential temperature from a mast as follows:*

> *40:   288 K*
> *100:   287 K*

What is Γ? Is the atmosphere stable, neutral or unstable? What would the value at 100 m be if the atmosphere was neutral?

4.26 *Calculate the wind profile in the case of a stable atmosphere, with $u_* = 0.4$, $z_0 = 0.15$, L = 250. How is this profile different the neutral version (i.e. assuming a neutral atmosphere).*

4.27 *Calculate the wind profile in the case of an unstable atmosphere, with $u_* = 0.4$, $z_0 = 0.15$ and L = −250. How is this profile different from the neutral version (i.e. assuming a neutral atmosphere).*

4.28 *What is the albedo of the Earth? Use Figure 4.12.*

4.29 *What would the temperature have been today, had the Earth had no atmosphere, and therefore no greenhouse effect?*

4.30 *Given a geostrophic wind of 12 m/s, and a roughness, z_0, of 0.75 m. Find the wind speed at 100 m's height at 60°N. (Remember A = 1.8 and B = 4.5).*

5

Local Flow

In the ideal world there would have been no need for this chapter! Because, in the ideal world, there would be one grand unified model that would cover the flow on all the scales we discussed in Chapter 2: global, synoptic, meso- and microscale, it would be very accurate and it would take no time to run. With such a tool, all we need to do would be to find the location on the globe, input the necessary data (wind and geographic) and we would – in no time – have the entire flow field of the site very accurately and at a very high resolution. This we could then use to find the optimal location of the wind turbines and estimate the long-term wind resource of the site, so we can make the necessary economic and technical decisions.

Unfortunately, we are quite far from this ideal world: we have no all-scale models. Most models are not as accurate as we would like them to be, and even in simplified and limited cases, the models can take some significant time to run. The consequence of this is that we need to develop models that represent the flow at a specific scale, one scale at the time. As described in Chapter 2, the flow at any given location is the result of the flows on all scales. However, for a given location what makes the difference between a good spot and a bad one (seen from a power-production point of view) is the local flows. So, in this chapter we will look at models that describe the *local scale* flow. This is done by identifying the effects that affect the local flow and then discuss the corresponding models, one local effect at the time. Referring to the description of scales in Table 2.2, we are going to look at the various types of flow on the *microscale*.

Exercise 5.1   *What are the characteristic length and time scales of the microscale?*

Again going back to a more ideal world, the flow on the micro scale should be modelled by one micro-scale model, there is after all just one flow. This is closer to the current real world; however, we are still not there. So in order to model the flow at this scale, we need to divide the flow into a number of different effects, and then model each of the effects, one at the time. Doing this will also enable us to understand, in our usual hand-waving physical way, what the characteristics of the different types of flow are, thereby again improving our 'feel' for the

Meteorology for Wind Energy: An Introduction, First Edition. Lars Landberg.
© 2016 John Wiley & Sons, Ltd. Published 2016 by John Wiley & Sons, Ltd.

flow at a given site. In the following, we will therefore go through each of the local effects, first trying to explain what the effect is, and then discuss the models that can be used to model this particular effect, starting with very simple ones, then moving on to briefly describing the more complex ones. The total local flow therefore needs to be build up – local effect by local effect – to get to the resulting flow we see at the site.

5.1   Local Effects

Since the publication of the European Wind Atlas (Troen and Petersen, 1989) the standard set of effects that affect the flow on the local scale has been as follows:

- Orography
- Roughness
- Obstacles

This made a lot of sense at the time, and these were the effects that could be modelled (by e.g. WAsP, Mortensen et al., 2007). However, for this discussion, I would like to add a fourth effect as well, namely the effect of:

- Thermally driven flows

The latter do not normally produce the strongest of winds, and they are not so easy to model. However, in many cases they can still alter the flow at a site quite a bit and by including them, we get a more complete picture of the flow on the micro scale.

In the following, each of these types of flow will be described in some detail. There is an underlying assumption that these effects are independent of each other – for the simple models that is true, but as the models get more and more complex, this assumed independence starts to break down, but it is still, nevertheless, useful to think about these effects one at the time.

Except for the simplest of cases, it is difficult to do calculations by hand, so there will not be as many exercises in this chapter as we have seen in some of the other chapters.

5.2   Orographic Forcing

Before we get into the details of this effect, it is important to realise that a very simple picture of the way the flow behaves when encountering the terrain, is that it follows it! This is quite a sweeping statement and it is often very wrong, but it can also very often help avoid some fundamental misunderstandings of what goes on.

In order to understand this effect, we shall start out with a very simple relation, called Bernoulli's equation.

One does not need to have seen many wind farms around the world before one realises that they have a tendency to be built on the tops of hills! The reason for this is due to the first of the local effects, the orographic forcing. This is a posh term borrowed from traditional meteorology, which basically means that the flow is forced by the terrain (or more specifically,

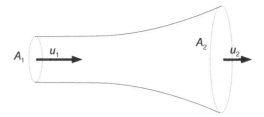

Figure 5.1 A schematic illustration of a flow tube and Bernoulli's equation (Equation 5.1)

the height variation of the terrain, which is what is called the orography) in a specific way due to the hills and valleys that it encounters. The forcing is such that the hills will speed up the flow and the valleys will speed/slow down the flow.

If you have a physical feeling for flows based on looking at water, for example, streams, rivers, pipes, and other such things, Bernoulli's equation might make perfect sense. If the area of the flow decreases, the water flows more quickly. However, if you think about air, it might actually not seem so obvious, since our daily experience with air is that it is compressible, that is, it compresses when we try to move it rather than flow like water. This is true of course; however on the scale of the atmospheric flow, the air in the atmosphere does act as if it is incompressible (due to something called the anelastic constraint), so at these scales, the Bernoulli equation also works for atmospheric flow.

The simplest way to describe this is to use a simple version of Bernoulli's equation, which states:

$$A_1 u_1 = A_2 u_2 \tag{5.1}$$

where A_i is the area of a cross-section of a flow tube, and u_i, the speed of the flow at location i. A flow tube is an imaginary construct, and to understand what it is, imagine that you trace a particle as it moves through the air. This trace is called a streamline, and a bundle of these streamlines is the flow tube – one characteristic of such a flow tube is that all the particles stay within the tube (as they need to, since it is their movement that defines the tube in the first place). An illustration of this can be found in Figure 5.1.

Imagine now that you look at the streamlines that flow over a hill, and a flow tube made out of these, Bernoulli's very simple equation then says that since a hill compresses the streamlines (i.e. A gets smaller) as the air flows over it, then the wind speed, u, will increase in order for the equation to hold. So on top of a hill the wind is sped up, hence all the wind turbines are on hills rather than in valleys.

Exercise 5.2 *Calculate the speed up of the flow in a tube where the area is reduced by 10%.*

Exercise 5.3 *What happens to the streamlines when the flow passes over a valley?*

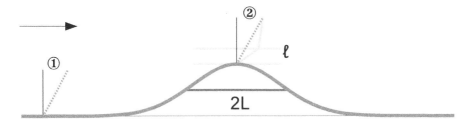

Figure 5.2 An illustration of the flow over a Gaussian-curve-shaped hill. The wind blows from left to right, and the profile shown at location 1 is the observed unaffected profile, called the upstream profile (dotted line). Note that the profile is plotted in the usual log–linear coordinate system. At the top of the hill, location 2, the observed profile is plotted again (solid line), together with the upstream profile (dotted line) for comparison. The hill has caused a general speed-up of the flow, with the maximum at height ℓ. The half width of the hill is indicated by L. The drawing is not to scale

Bio 6: Daniel Bernoulli[1]
1700–1782
The Netherlands
Dutch mathematician, born out of a prominent family of mathematicians. Applied mathematics to mechanics, especially fluid mechanics, and did pioneering work in probability and statistics. Did most of his work in Switzerland.

5.2.1 Analytical Models

We now have a very simple first idea of how the speed up/down over hills and valleys works. The next step is to look at a slightly more complicated model, but only slightly more so. This model has been developed for a single and two-dimensional hill (two dimensional in the sense that it stretches to infinity in the third dimension/sideways), shaped like a Gaussian curve (i.e. a bell-shaped ridge).[2]

Looking at Figure 5.2, two characteristic length scales are defined, L and ℓ. L is defined as a 'typical length' of the hill, and what that means is that L characterises the length of the

---

[1] Figure source: "Daniel Bernoulli 001" by Johann Jakob Haid – Here, https://commons.wikimedia.org/wiki/File: Daniel_Bernoulli_001.jpg#/media/File:Daniel_Bernoulli_001.jpg [Public domain], from Wikimedia Commons

[2] This part is very much inspired by the work done around the Askervein experiment (Taylor and Teunissen, 1987) and the writings in the European Wind Atlas, mentioned above, and its predecessor (Jensen et al., 1984).

hill in some standardised way – for Gaussian hills, L is the half-width of the Gaussian curve (i.e. the width of the curve when the value (height) is half the maximum). ℓ relates to the flow over the hill more than the dimensions of the hill itself, and is defined as the height where the speed-up effect of the hill starts to decrease, indicated by a kink in the vertical profile of wind speed.

This might sound complicated, but if we have a Gaussian hill, then we can approximate the maximum speed up, ΔS, and ℓ using the following two equations:

$$\Delta S \approx 2\frac{\ell}{L} \tag{5.2}$$

and

$$\ell \approx 0.3z_0 \left(\frac{L}{z_0}\right)^{0.67} \tag{5.3}$$

where z_0, as always, is the roughness. Just to reiterate, this means that one can calculate the fundamentals of the flow over a simple hill, just by knowing its dimensions.

There is one further property of the flow that can be explained and approximated by this simple model and that is at a height of $2L$ the speed-up effect of the hill has vanished (this is of course not entirely true, but the effect is now so small that this works as a good approximation).

Let us try to look at an exercise to make this clearer:

Exercise 5.4 *Imagine you have a bell-shaped hill that can be approximated by a Gaussian function. The height is 250 m, the half-width is 500 m and the roughness of the surface is 0.1 m. Draw the hill. What is the speed-up at its maximum, and at which height? At what height has the speed-up effect of the hill vanished?*

The hill is drawn in Figure 5.3 and in order to calculate the maximum speed up, we first need to calculate ℓ:

$$\ell \approx 0.3z_0 \left(\frac{L}{z_0}\right)^{0.67} = 0.3 \cdot 0.1 \left(\frac{500}{0.1}\right)^{0.67} = 9.0 \tag{5.4}$$

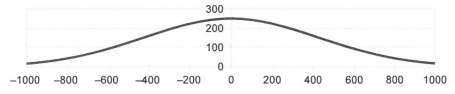

Figure 5.3   A plot of the Gaussian hill discussed in Exercise 5.4

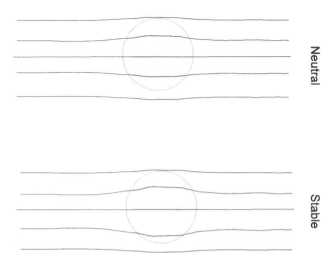

Figure 5.4 Flow around a hill seen from above. Top shows the flow under neutral conditions, bottom under stable conditions. It might not be very easy to see, but the streamlines for the stable case are going more around the hill than in the neutral case

so we find the maximum speed up not too far from the surface at 9 m, using ℓ to calculate ΔS, we get:

$$\Delta S \approx 2\frac{\ell}{L} = 2\frac{9.0}{500} = 0.04 \tag{5.5}$$

meaning a 4% speed-up at the maximum. Not the most dramatic of speed-ups, but this actually means that this simple expression is probably not too far off. The speed-up effect of the hill vanishes at $2L$, that is, at $2 \cdot 500 = 1000$ m.

5.2.1.1 Direction Dependence

The direction of the wind is also affected by the terrain; unfortunately, there are no simple expressions for this. In the following I will cover the two most basic facts about direction: how the flow behaves over and around a hill, and how this flow is influenced by stability.

In Figure 5.4 the flow over the same hill is illustrated for two different stability cases: neutral and stable. In the neutral case the flow does go around the hill, but the direction is not significantly changed. In the stable case, the flow is 'pushed' from above, and the flow is forced to go more around the hill.

Box 13   Mass-consistent models

Mass-consistent models are not used so often any more, and the lack of accuracy and complexity are a clear drawback, however, you might come across them when quick and dirty calculations are required, so in this box we will go through them briefly.

The mass-consistent models are used to calculate the wind field based on observations at the site, the more observations the better. Taking the terrain, the roughness and the observations (of wind speed and direction and in some cases also temperature) into account and assuming that mass is conserved (hence the name), the wind field covering the entire site can then be calculated. Only mass conservation is taken into account, so all the dynamic parts are missed out, hence the limited accuracy. The one main advantage is that these types of models are fast.

The main equation that governs these types of models is the equation for conservation of mass, also called the continuity equation:

$$\frac{\partial \rho}{\partial t} + \nabla \cdot (\rho \mathbf{u}) = 0$$

which for incompressible flow (meaning that the density, ρ, is constant), like the atmosphere, can be reduced to:

$$\nabla \cdot \mathbf{u} = \frac{\partial u}{\partial x} + \frac{\partial v}{\partial y} + \frac{\partial w}{\partial z} = 0,$$

solving this, we get the flow field over the site.

5.2.2 Attached Flow: Flow in Simple Terrain

As we increase the complexity of the modelling, we move further and further away from being able to explain things by simple equations and closer and closer to having to solve the full set of flow equations using numerical methods. I will not explain these equations here, but have given an introduction in Box 14.

Box 14 The Navier–Stokes equations
{*You can skip this box if you wish.*}

The Navier–Stokes equation for incompressible flow, states the following:

$$\frac{\partial \mathbf{u}}{\partial t} + (\mathbf{u} \cdot \nabla)\mathbf{u} = -\frac{1}{\rho}\nabla p + \nu\nabla^2\mathbf{u} + \frac{1}{\rho}\mathbf{F}$$

First of all, do not forget that this is just Newton's second law of motion, that is, acceleration on the one side and the forces on the other. If you refer back to Chapter 2 you will actually be able to recognise quite a few of the terms, but let us start with the left-hand side, called the convective terms. These two terms are just saying that the equation is describing the acceleration and convection when following an air parcel. Even though you are not familiar with the various operators you might suspect that the second of the two terms involves something which is non-linear (if you don't, don't worry!): $u \cdot \nabla u$. This is the term that causes the difficulties when trying to solve the Navier–Stokes equations, and I will return to it in Box 16.

On the right-hand side all the forces are lined up and we see three terms. The first one, which is an old friend, the pressure force, is followed by a term we have not seen before, called the viscosity/diffusion term, which we will return to shortly, and finally a catch-all term, $\frac{1}{\rho}\mathbf{F}$, summing all the other forces acting on the flow (like gravity). The middle term, the viscosity/diffusion term is expressing how momentum is diffused, that is, how fast can momentum be transported in the fluid. v is the so-called kinematic viscosity, and is given by the dynamic viscosity, μ, divided by the density, ρ (again see Box 16).

The above equation is, like Newton's second law of motion expressing that momentum is conserved. The full set of Navier–Stokes equations is comprised of two other equations as well, one expressing the conservation of mass, which we saw in the box on mass-consistent models, and one expressing the conservation of energy.

It should be noted, that without significant simplifications the Navier–Stokes equations cannot be solved analytically, so the only tool available is the set of models called computational fluid dynamics (CFD) codes, which we will learn about shortly.

The first set of models that we encounter are the models that are able to represent the so-called attached flow. The easiest way to explain the types of terrain that cause this type of flow, is to imagine that the terrain is flat with just very very small bumps. You don't find that type of terrain in many real places, maybe with the exception of my mother country Denmark! However, it is quite useful to consider these kinds of terrains, since the number of other terrain types that can be approximated by this assumption is actually quite large. Attached flow means that the streamlines follow the terrain and therefore there is no separation (see next section). Given that the flow is attached, it is possible to cut quite a few corners in the modelling, and thereby simplify the calculations, and making them faster too.

One group of models which is able to represent the attached flow is called *linearised models*, where the 'linear' refers to the fact that some of the terms in the Navier–Stokes equations (see Box 14) are changed from being non-linear to approximate expressions that are linear. You might realise that doing this kind of simplification of course severely limits the flow types where the models apply. The trick is to look at the terrain and then judge if a simple model can actually be used in that particular case, and when looking at the results, take the limitations into account.

The classical set of models that belong to the linearised flow models are the ones of the Jackson–Hunt type (see Jackson and Hunt, 1975). These models have been incorporated in various flow model packages, the most widely used being the WAsP model (Mortensen et al., 2007)[3] mentioned earlier. Other versions of the model are MS3DJH (Salmon and Walmsley, 1986) and similar codes.

---

[3] Disclosure: In my previous job I was heavily involved with this programme.

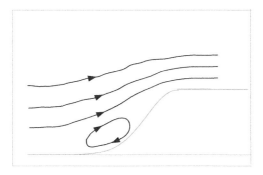

Figure 5.5 Illustration of detached flow caused by an escarpment. The illustration is very simplified as you will see solving Exercise 5.5

5.2.3 Detached Flow: Flow in Complex Terrain

As the terrain gets steeper and steeper something happens to the flow: it *separates*. This is a binary process at one point the flow is attached, increase the steepness just a bit, and it separates. As illustrated in Figure 5.5, separated flow means that, as the word says, the streamlines near the surface stop to follow the terrain, and not only that, when the flow is separated a *separation bubble* is formed, where the flow near the surface flows in the opposite direction of the overall flow.

One way of illustrating this is of course by doing it as I have done in Figure 5.5; however, a much better way is to show it as a movie. There are a lot of these on various sites on the internet, and they change quite often, so please do the following exercise now.

Exercise 5.5   *Search for videos of detached flow.*

As you can see, this is a fascinating phenomenon to look at. And by studying a few – if not many of these – movies, you will get a good feel of how separated flow looks, and possibly also under which conditions separation occurs. The effect wind-speed-wise is that the streamlines are less compressed than they would have been, had the flow not been detached (this is of course something you need to imagine, since in reality, the flow will always separate under these conditions).

Exercise 5.6   *What does less compression of the streamlines mean? Please think about this before you proceed.*

Less compression of the streamlines means less speed up of the flow, that is, relatively lower wind speeds. What happens inside the separation bubble is of course far from simple. The bubble is also called the recirculation bubble (in the recirculation zone), alluding to the fact that the flow goes the opposite way, and as a consequence the concept of a well-defined flow tube breaks down.

Box 15 Onset of separated flow

An important question to ask when modelling separated flow is: at which steepness of the terrain does the flow separate? This is an empirical question that can be answered, based on direct measurements, wind tunnel studies and other observations. Recently, even very advanced flow models (see Section 5.2.4) have been used to shed light on this question. The current thinking is that there is a difference in this onset between when the flow goes up the hill (the upstream side) and when it goes down the hill (the lee side). The steepness required for the upward-going flow to separate has been found to be much steeper than on the lee side. Some authors report numbers as high as 0.92 (Ferreira et al., 1995), and many others struggle even to detect any separation. The downward/lee-side steepness has been found to be around 0.3, and Wood (1995) has developed an expression for the onset of separation that only depends on a characteristic of the dimensions of the hill and the roughness, much like we saw with the simple speed-up model.

Due to the corners cut in the linear models discussed above, linear models are not able to model separated flow. Using a classic linear model to calculate separated flow would result in modelled flow that followed the terrain, no matter the steepness, compressing streamlines more and more, whereby increasing the wind speed more and more. As we have just seen, this would mean that the speed up on for example, a steep hill would be over-predicted by such a model, and seriously over-predicted in the case of very steep terrain. Very steep terrain is called *complex terrain*.

It is a little bit depressing, but also an illustration of how difficult the problem is, that ever since I started working in wind energy (25+ years ago), one of the items highest on the research agenda has been to solve the problem of flow in complex terrain! The issue has been that even the most advanced models have had a very difficult time beating the simple models we discussed in the previous section, meaning that the predictions of both types of models could be very inaccurate.

Different ways of making the simple models more accurate, even when they are used outside of their so-called operating envelope, for example, running them in too steep terrain, have been developed. I will mention just one here, one which I have been involved in: the Ruggedness Index or RIX (Bowen and Mortensen, 1996). The index is a way of correcting the model based on the terrain steepness (or rather the difference in the steepness of the terrain). This index and other different ways of correcting the output from linearised models, does indeed improve things a bit, and in some cases quite a bit, but again, of course the art is to know when the improvement is real and when it is not.

5.2.4   More Advanced Models for Flow in Complex Terrain

As models have got better and better, but especially as computers have got faster and more powerful in other aspects, too, we have recently seen significant improvements in the modelling of flow in complex terrain. It can be argued exactly when the much-hoped-for point where the advanced models actually outperform the simple linear models has occurred, but – as opposed

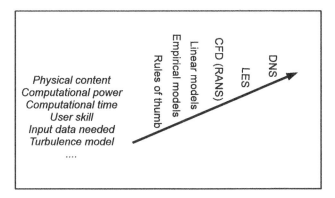

Figure 5.6 The wide range of flow models, from the very simple to the very advanced. The *y*-axis represents the various dimensions of complexity and they all increase as the overall complexity increases. The acronyms are explained in the text

to the 'old days' – more often than not do the advanced models beat the simple ones. I need to qualify this statement a little bit, since the models are now so advanced that it is not just a matter of installing them, running them and then thinking that the results are right (which is almost how the simpler models can be run) – as the models get more advanced, the user needs to be so too, and not understanding how a model works is no longer an option. Please refer to Section 8.3 for a much more detailed discussion of this aspect.

If you remember from the section on the linear models, the models were simpler exactly because the Navier–Stokes equations were linearised. So, by not linearising the equations for example, going back to the full set of terms in the equation, we end up with the more advanced models (from RANS and upwards in Figure 5.6).

These models are called computational fluid dynamics (CFD) models. It is of course a bit misleading to use this term, since the linearised models also compute the dynamics of the fluid, just at a less complex level. However that might be, CFD models in wind energy (and most other fields) refer typically only to the more advanced models, and specifically the term is used for the Reynolds-averaged Navier–Stokes (RANS) models (see Section 6.2 and Box 14).

Given the scope of this book, I will not go into too much detail here, but one of the important elements of the advanced models (up to a certain level) is turbulent closure models (which is actually also a simplification), that is, how do we make models that as accurately as possible represent the turbulent part of the flow? Please refer to Box 16, if you want to know a little bit more.

Box 16 Turbulent closure
{*you can easily skip this box*}

We will discuss turbulence a lot more in Chapter 6; however, since how turbulence is modelled is such an important part of modelling separating and separated flow, I shall discuss the so-called turbulent closure schemes briefly, very briefly actually. You can find a bit more in Box 18.

As indicated in the box on Navier–Stokes equations (Box 14), the convection term:

$$u \cdot \nabla u$$

which after Reynolds averaging (see Section 6.2) can be written as the product of the turbulent parts of the flow, for example:

$$u'v'$$

is a non-linear term, which means that it is difficult to work with it mathematically. What turbulent closure models do is that they try to 'close' this problem by relating the u', etc. to the mean flow (the u's). This is done through a quantity called the eddy viscosity (i.e. a viscosity for the turbulent eddies). This quantity in turn can be related (in some formulations) to another quantity called the mixing length, which has to do with the physical dimensions of the flow.

There are many different closure schemes employed, of varying complexity, but to name-drop two:

- $k - \epsilon$, a two equation model
- Reynolds stress equation models, which are the most complex classic models.

As illustrated in Figure 5.6 the different flow models we have discussed this far sit on the spectrum from low complexity to high complexity, from simple to advanced, from requiring limited computational resources, to requiring all you can get, etc.

At the far end of the spectrum we find two very advanced and computationally demanding models, the large eddy simulation (LES, which as the name says, numerically and explicitly resolves the largest eddies, which also carry the highest amounts of energy) and the direct numerical simulation (DNS) models. The problem both of these models try to address is how to model the turbulent part of the flow. As we saw in the box on turbulent closure (Box 16) this is in general what the models with the more advanced models try to improve on. I will not go into any detail at all about these two more advanced types of CFD model variants, but since the resulting calculated flow fields are quite nice and illustrative (compare to the movies you found in Exercise 5.5) I will show one example here: a flow simulation using an LES model, see Figure 5.7. As you can see, the increased complexity and sophistication of the modelling results in very detailed and high-resolution flow fields.

To finish this part on flow modelling it is useful to be able to ask some relevant questions 'of' the flow models, that is, when you are presented with the output of a flow model, these are then the questions to ask. The first most general one is: is this the right model to use for the job? The other is: exactly which model has been used, that is, version number, etc. Another question to ask is what values for the parameters have been used, for example, the ones in the turbulent closure model. Other questions should focus on the grid resolution and the time steps used. There are also some more philosophical questions to ask, but they will be covered in Chapter 8.

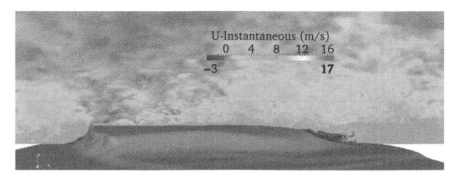

Figure 5.7 An example of the output from a LES model. Here a simulation of the flow over the Bolund peninsula. The shades of grey represent the wind speed, dark: high, light: low. See Diebold et al. (2013) for further information. Source: © DTU Wind, Denmark. Reproduced with permission of DTU Wind, Denmark

5.3 Roughness

Roughness, and the effect of roughness on the local flow, actually crosses into most of the other models in one way or another – even the simple ones (as e.g. illustrated in Equation 5.3). Since we have already discussed vertical wind profiles (in Chapter 4) where roughness is an essential part of how these profiles look, and since we have also already discussed internal boundary layers (IBL), we are very well prepared to discuss the effect of roughness on the local flow. In the simple way of discussing the various flow effects, these are, actually, the only two effects that are normally included when discussing the effects of roughness.

This also means that there are no further models that are required when addressing the effects of roughness. However, since roughness is so important in many, if not all, of the other models for local flow effects, estimating the roughness becomes quite important.

Exercise 5.7 *(Repetition) How did the expression of the logarithmic profile look, and how was it dependent on the roughness?*

Exercise 5.8 *Imagine that you are not entirely sure what the roughness of your site is. You guess that it is probably between 3 and 10 cm. You have been given some measurement data at 30 m, and you are asked to estimate the wind at hub height (100 m). The mean wind you have been given is 7.5 m/s. What is the span of the hub height wind speeds given these two roughnesses?*

If you do the calculations (see Appendix B, under Chapter 5), then you will see that the span is actually quite large (from 8.8 to 9.3 m/s or 6%). This stresses the fact that estimating the roughness is important, and going from mean winds to power production makes this even more important (due to the fact that, as we have seen, power goes as the wind speed squared or even cubed).

There is one issue that can be a bit difficult to get your head around at first and that is not to confuse the terrain with the roughness. As an example, consider a rugged mountain

Table 5.1 The relation between landscape type and roughness. Based on the classification in the European Wind Atlas (Troen and Petersen, 1989). Only a few of the classes are shown here

Landscape type	Water	Snow	Grass	Farmland	Forest	City
Roughness (m)	$2 \cdot 10^{-4}$	10^{-3}	0.03	0.1	0.8	1.0

landscape which can easily have a lower roughness than a forest on a completely flat piece of land. So, when estimating the roughness, it is important to, so to speak, 'take the air out of the landscape', that is, just using what is *on* the surface, as the basis for the roughness calculation. The reason for this is that – to a first approximation – the flow follows the terrain (orography).

To aide in the roughness estimation, there are various tables that relate a specific landscape type to a roughness. The one that I will show here is the classification from the European Wind Atlas, see Table 5.1. As just noted above, remember to only look at the roughness, not the orography.

The landscape types in the table cover many of the cases you might come across, but it does not cover them all, so you will need to do a bit of mental roughness interpolation to come up with the correct value for the roughness of your site in the cases where the landscape type does not fit in one of the ones in the table. It is important to stress that roughness estimation often is quite subjective, which of course is not ideal, given the importance it plays in the calculated power. Often the roughness needs to be estimated based on satellite pictures, aerial photos or photos taken at and near the site. The only real case when roughness of one type of landscape can be *measured* is when we have an infinite area of it! Often a little less than infinite will do, of course, but we are still talking about large areas of homogeneous upstream landscape for this to work.

Exercise 5.9 *How would you measure the roughness value of an infinite, homogeneous landscape?*

As you can see from the table, one single value has been assigned to the roughness of water, $2 \cdot 10^{-4}$ m. This is fine when looking at climatological, that is, longer term effects (say many years). However, if we want to look at it more physically, there is – as you might be able to imagine, comparing a calm sea surface, to that of a stormy day – a dependence on wind speed. This is typically expressed through u_*. The traditional expression for the roughness of a water body, z_0^w, is the one by Charnock (1955) which says:

$$z_0^w = a_c u_*^2 / g \tag{5.6}$$

where $a_c = 0.015$ (approximately), u_* the friction velocity and $g = 9.81$ m/s^2 the gravitational constant. The equation works from around 3 m/s and up. As you can see from this equation, the higher the wind, the higher the roughness, as we would expect.

A final complication (sorry!), when estimating roughness, is that it varies with the seasons! Think of a field covered in snow during the winter and covered with for example, corn during the summer, and harvested in autumn, the signature on the map will be exactly the same (most likely some shade of green), but the roughness will vary immensely, by orders of magnitude.

Estimating the roughness based on a picture from the site (which always will be from just one of the seasons) will therefore clearly not do the trick. First you will need to know what the purpose of the roughness estimation is, and if it is for example, for a climatological estimate, you will need to let the seasons fly by in your mind's eye, and average the roughness as you go along. Adding, unfortunately, further complexity and subjectivity to the problem, thereby also introducing more uncertainty.

Exercise 5.10 *Estimate the roughness of a field in the following three cases: covered in snow, corn and just harvested.*

5.4   Obstacles

An obstacle is typically some sort of building or shelter belt (a row of trees is a good example of a shelter belt). The effect of an obstacle on the flow is to reduce the wind speed and increase the turbulence, especially downwind. The closer to the obstacle, the more reduced the speed. In general terms, the flow is also more complicated close to the obstacle than further away. Just as we saw in the case of the hills, there is also an effect (a reduction) in front of the obstacle, but this is generally in a much smaller area, and in most of the simple models this is rarely taken into account.

The modelling of flow around obstacles has lived a life in and out of the limelight, and is currently coming back on to the stage. In the 'old' days, the modelling of obstacles was very important, since much of the available data was from either airports or light houses taken from instruments at low heights. In both of these cases, the flow was affected strongly by obstacles, think houses and hangars, nearby the mast. Understanding the effect on the flow of these was therefore very important. As we then got more and more longer-term, on-site data at much higher heights from wind farm sites, which were mainly free of any obstacles, this type of modelling was not so important, and for many years this was the case. In relatively recent times, however, wind turbines and sometimes entire wind farms have started to be build in harbour/industrial, even urban areas (see e.g. Figure 5.8), and in these cases the flow – this time at the turbine location – is again affected by obstacles.

There is a very nice (if you think that equations can be nice!) theory that predicts how an obstacle affects the flow. It has one peculiar aspect which we will discuss first: it normalises lengths and heights with the height of the obstacle. This is a bit backward at first sight, but by doing so the theory simplifies very nicely. So, let us see what this normalisation means.

Exercise 5.11 *Imagine that the flow you are looking at is affected by an obstacle (think of a house) which is 5 m high. Your point of interest (think of either a mast or a wind turbine) is 600 m away and has a height of 50 m. How many obstacle heights is the point of interest away from the obstacle? What is the height of the point of interest in obstacle heights?*

(Please solve this before moving on).

Figure 5.8 Spot the wind turbine! A wind turbine in an industrial/residential setting. The flow that reaches the turbine from almost any direction will be affected by obstacles. In all fairness I think this turbine is more meant as PR/a landmark than as an optimally placed and producing wind turbine. Source: © Landberg 2014

Since the height of the obstacle, h, is 5 m, the new 'yardstick' is 5 m, so a distance of 600 m, turns into $600/5 = 120h$, that is, 120 obstacle heights and a height of 50 m turns into $50/5 = 10h$, that is, 10 obstacle heights.

We now understand the normalisation with obstacle height, so the equation that describes the effect on the wind speed can be shown. The one I shall use here is that of Perera (1981), and it states:

$$\tilde{u} = 9.75(1 - P)\eta\frac{h}{x}\exp(-0.67\eta^{1.5}) \qquad (5.7)$$

where $\tilde{u}$ is the normalised velocity deficit, P is the porosity, that is how open (porous) the obstacle is, $P = 0$ means solid, and $P = 1$ means thin air. η is a variable dependent on z, h, and z_0 (and strictly speaking also the displacement height) and h, the height of the obstacle, as we saw before.

As you can see, knowing the obstacle height, collapses all different obstacles into one equation. You might not think that this is an extremely simple or even nice equation, but looking at it in Figure 5.9 you can see the resulting reduction; there are a few interesting aspects of the equation. The effect of an obstacle can be quite dramatic, right behind it (but outside the zone where the model is not valid) we find reductions above 50%, and even quite far away (say 30 obstacle heights away) and quite high up (say two obstacle heights) the reduction is still around 15%, and note this is the reduction of the wind speed.

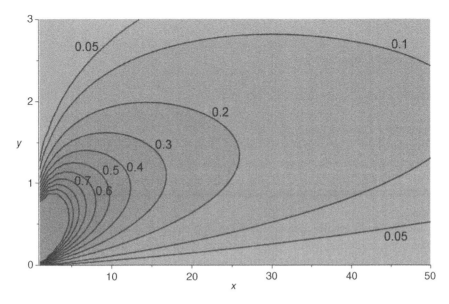

Figure 5.9 The reduction of the free-stream wind caused by an obstacle with height h, c.f. Equation 5.7. Along the x-axis is plotted the horizontal distance from the obstacle normalised by h, and along the y-axis the height above ground level, also normalised by h. Each contour line is labelled with the reduction. The results of the model are not reliable very near the obstacle

We briefly need to return to porosity, since the two issues we had with roughness repeat themselves here too: subjectivity and seasonal variability. It is difficult to measure the porosity directly from either a photo or a description, which introduces the subjectivity. If you imagine, say a row of trees, they change their characteristics a lot over the seasons, thereby changing the porosity. If the obstacle is a building, there is no subjectivity (at least in this simple model) and most likely no seasonal variation (!).

5.5 Thermally Driven Flows

As mentioned in the introduction to this chapter, thermally driven flows are not traditionally included in what is called local effects. However, as I also mentioned above, by including them here, we get a more complete picture of which phenomena influence the local flow.

The two fundamental causes behind thermally driven flows are the – simple – physical facts that (relatively) hot air rises and (relatively) cold air sinks; and that light air rises and heavy air sinks. We need to distinguish between these two because air can be heavy for other reasons than temperature differences.

In the following, two types of thermally driven flows are described:

- Sea/land breezes
- Ana-/katabatic flow

5.5.1 Sea/Land Breezes

Sea breezes occur near and at coastlines, and are caused by the fact that during the day the land will heat quicker than the sea.

Exercise 5.12   *Why does land heat quicker than the sea?*

(Think about this for a short while, please)

This is because the heat capacity of water is much higher than that of land. The heat capacity of a substance is a measure of how much heat is needed to raise the temperature of the substance 1°C. For water it is 4186 J/kg and for dry soil 800 J/kg, that is, a difference of more than a factor of five. Meaning that, when the Sun shines, the soil heats up more quickly than the water.

The hotter land surface will cause the air to rise and the air will sink over the cooler water. To preserve mass, a resulting flow, going from the water to the land near the surface and opposite aloft will develop (see Figure 5.10). A person standing at the shore will therefore feel a light wind coming from the sea. The wind speed is typically only a few metres per second, so calling this a breeze is appropriate.

My favourite example of the effect and possible consequences of a sea breeze is the tropical island sketched in Figure 5.11. The sea breeze is perpendicular to the coastline, so at the mast location the wind is westerly, that is, coming from the west. During the day, however, where we want to build our wind farm it will be northerly. This would make the measured directional distribution (the wind rose) completely wrong (if it is used as being representative at the wind farm location), resulting in a potentially completely wrong layout of the wind farm. Note that this is just an example to highlight an important point, the sea breezes are often quite light in reality.

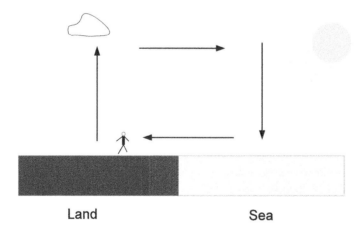

Figure 5.10   An illustration of a sea breeze. As the land gets (relatively) hotter, the air starts to rise over it, and descent over the water, the flow near the surface goes from the water to the land, and opposite aloft

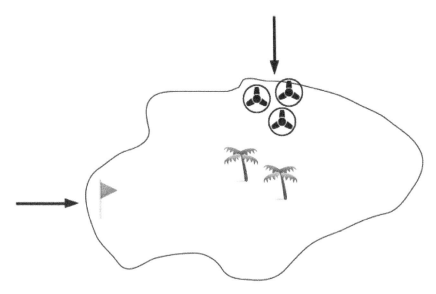

Figure 5.11 Sea breezes on a tropical island. The flag marks the met mast, and the rotors the wind farm. The arrows indicate the direction of the sea-breeze flow

Leaving now the tropical island and returning to the coastline please think about the following question:

Exercise 5.13 *What happens at night at a coastline?*

At night, there is no external heating, but due to the difference in heat capacity, the land cools quicker than the water, meaning that now the air will sink over the relatively cooler land and rise over the water. The counter flow will result in the wind flowing offshore near the surface (and onshore aloft). A land breeze has developed. The winds are very weak indeed in this case and do probably not play any role in wind energy.

5.5.2 Ana-/Katabatic Winds

The second type of thermally driven flows that we are going to discuss here is the Ana- and Katabatic winds. You might recognise the 'an' and the 'kat' from batteries (the anode and the cathode) and the two words mean 'upwards' and 'downwards', respectively.

As illustrated in Figure 5.12, anabatic flow can occur when the air is heated (relative to its surroundings) at for example, a mountain side. Hot air rises, and as it cools (due to the fact that it is not near the hot surface anymore), it starts to descent. Thereby a – gentle – circulation is formed.

In the case of a katabatic wind (see Figure 5.13), which often occurs at night, the air parcels for example, on top of a mountain, are cooled as the atmosphere radiates heat to the surface (which in turn radiates heat to space, as you might remember from Chapter 2), and since cooler

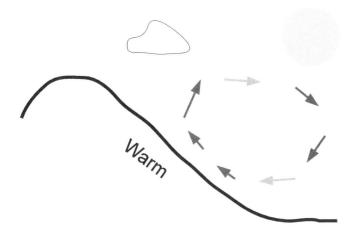

Figure 5.12   Anabatic flow. The Sun heats the air parcels near the mountain side, they get lighter and start to ascend, creating a (weak) flow circulation. As the air rises, the water vapour condenses, sometimes creating clouds above the flow

air is heavier than hotter air, the cooled air starts to descend, and as it gets cooler and cooler it also gets heavier and heavier and accelerates, resulting in, in some cases, very strong winds.

Due to their strength and destructive (and sometimes annoying) nature, many anabatic and katabatic winds have been given local names. Some (in)famous examples are the Santa Anna (California, United States, strong, extremely dry down-slope winds), Piteraq (Greenland, meaning 'that which attacks you', 50 to 80 m/s wind), Chinook and the Föhn, (down-sloping, dry, warm air). In the Mediterranean, we find the Mistral (cold, northwesterly wind, up to 25 m/s), Bora, and the Tramontana. Oroshi (Kanto Plain, Japan) and Williwaw (cold, dense air from the snow and ice fields of coastal mountain areas around Alaska) are two other examples.

The effect on the wind resource of most of these winds is quite limited, but knowing about them enables you to explain phenomena at your site you might not otherwise be able to and also understand the extreme winds that the turbines at your site might be exposed to.

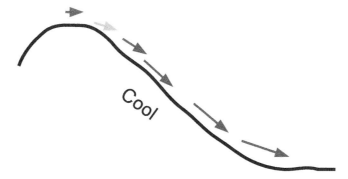

Figure 5.13   Katabatic flow. As the warmer air gets cooled by the (relatively) cool surface, it gets heavier and heavier and the air parcels accelerate downhill

Not many simple models exist that can model these thermally driven flows. We need to get to some level of sophistication to find such models. The type of model that I will briefly described here is the mesoscale models.

Exercise 5.14 *What are the typical length and time scales of the mesoscale?*

These models are not only able to model orographically forced flows, they also take roughness into account, and most importantly in this context, they can also model thermally driven flows. This is done through adding a few terms in the flow equation (see the Navier–Stokes box), and by introducing a new equation for temperature. There used to be a lot of different mesoscale models in use for wind energy purposes, but over the past many years this has converged to almost only one model, the weather research and forecasting model (WRF, Skamarock et al., 2008). This model is developed by NCAR (National Center for Atmospheric Research) and there are literally thousands of scientists working on it. This is of course a huge advantage, since many applications are being explored and the model is under continuous development, it has also the disadvantage, however, that when people say 'WRF' it is not so extremely clear which version, what parametrisations and which sub-models are actually used. So, as we will discuss further in Chapter 8, make sure you know the model you see the results of.

5.6   Effect of Stability

As we have learnt in the chapter on the wind profile (Chapter 4), many models assume that the atmosphere is neutral. This simplifies things, and often gives reasonable results (but of course sometimes completely wrong ones, too), but as the sophistication of the models increases, it is not possible to ignore the effect of stability. I have already mentioned the effect in quite a few cases above, and I will not go into any further details here, but just highlight the fact that the models for orographic forcing, in particular the more advanced models (the CFD/RANS models), to resolve all the possible flows at the site, also need to be able to model the effect of stability. The most difficult part of the stability range is the stable flows, and currently (2015), successful attempts are being made to model this (Bleeg et al., 2015). Another example is the models for IBLs: despite the fact that they are not very sophisticated, taking stability into account also improves them quite a bit in some cases (see Elliott, 1958, for some early thoughts on this). The models for thermally driven flows do of course need to be able to model the temperature equations, and thereby stability. So, to summarise this part about stability, in many cases the neutral models are perfectly adequate for modelling the flow. However, in some cases stability does need to be taken into account, too. Again, knowing when is what makes the whole difference.

5.7   Summary

This chapter (along with the one on profiles) is probably one of the more important ones in this book. Almost no matter how you are involved in wind energy, the flow at your site plays a very important role.

Table 5.2 Overview of the various local effects, and the ability of different numerical models to model them. $\sqrt{}$ means that the specific model is able to model the specific local effect, ($\sqrt{}$) that it can to some extend and in some cases, and X that it cannot. This is of course my take on this, and as the models develop, the 'scoring' will change

Model	Orography	Roughness	Obstacles	Thermal flow
Rule of thumb	$\sqrt{}$	$\sqrt{}$	X	X
Linear models	$\sqrt{}$	$\sqrt{}$	$\sqrt{}$	X
CFD-type models	$\sqrt{}$	$\sqrt{}$	($\sqrt{}$)	($\sqrt{}$)
Mesoscale	$\sqrt{}$	$\sqrt{}$	X	$\sqrt{}$

In this chapter we covered the following local effects:

- Orography
- Roughness
- Obstacles and
- Thermally driven flows.

We discussed what the causes were and also how to model the effects. We covered quite a lot of different models and to get an overview, they are all listed in Table 5.2. In the table I have also listed all the local effects, and for each combination I have specified whether the model in question is able to model the effect. As you can see, not one model is able to model everything, and a $\sqrt{}$ in the table means something different for a rule-of-thumb model than for a CFD-type model. One often used solution to this is to *nest* the models, that is, to put one model inside another; we will return to this in Chapter 8.

Exercises

5.15 *(Repetition) What was the rule of thumb for the rate of change of the height of the IBL?*

5.16 *Make a simple drawing of the wind rose at the mast and at the wind turbine locations for the tropical island in Figure 5.11.*

5.17 *Calculate the speed down of the flow in a tube where the area is increased by 15%.*

5.18 *Estimate the reduction of the wind speed 400 m behind and 15 m above ground level caused by a 10 m tall obstacle.*

5.19 *Estimate the roughness of the sea, when the friction velocity (u_*) is 0.6 m/s. Use Charnock's relation.*

5.20 *Estimate the roughness of a suburban area.*

6

Turbulence

When we discuss turbulence, it is important that we change the way we think about how we can understand and thereby model a physical phenomenon. Up until now (with only a few exceptions) we have been able to calculate more or less the direct consequence of a given phenomenon, we call this a deterministic system (see the chapter on modelling, Chapter 8, for more on this). When talking about turbulence, this is no longer possible, however, and we need to express our knowledge in different ways, like formulating statements about the statistical properties of turbulence, for example, through so-called spectra and classic time-series analysis. In fact, turbulence is so complicated, that it has both made the 'List of unsolved problems in physics' (Ginzburg, 2001), and it is also one of the seven Millennium Prize Problems (Clay Mathematics Institute, 2014) at par with some other very hairy problems. To top the list, Nobel Laureate Richard Feynman described turbulence as 'the most important unsolved problem of classical physics' (USA Today, 2014).

Turbulence is an important and difficult topic and there is, as just indicated above, unfortunately, nothing basic about it, but to set the scene here is a poem that tries to explain how it works[1]:

> Big whorls have little whorls
>
> That feed on their velocity,
>
> And little whorls have lesser whorls
>
> And so on to viscosity.
>
> *Lewis F Richardson, 1920 (Richardson, 2007)*

To put a few more, less poetic, words on this, this simple description of the so-called turbulent energy cascade says that the energy is introduced to the system via the very big eddies (another name for the 'whorls'), which in turn break up/generate smaller eddies through their motion,

---

[1] Source: © Cambridge University Press. Reproduced with permission of Cambridge University Press.

Meteorology for Wind Energy: An Introduction, First Edition. Lars Landberg.
© 2016 John Wiley & Sons, Ltd. Published 2016 by John Wiley & Sons, Ltd.

Figure 6.1 Turbulent and laminar flow on the river Rhine. The video this still is taken from can be found on the website, QR code in margin. Source: © Landberg 2014

and then smaller eddies generate even smaller ones, this continues until the eddies are so small that their scales are comparable to the molecular scale, at which point the cascaded energy ends up as molecular motion, that is, viscosity and heat.

There is a more scientific definition as well (see Tennekes and Lumley, 1983) and it includes a number of characteristics like irregularity, and that it is three-dimensional in nature. It is also important to note that turbulence is a characteristic of the flow and not the fluid.

To start out on the more stringent definition of turbulence, imagine flow in a pipe, like a see-through water pipe. The diameter of the pipe is L and the velocity of the fluid in the pipe is U. The viscosity of the fluid is v. Imagine next that we gradually increase the velocity. At first the flow is laminar, that is, it flows along straight lines; as the velocity increases, the flow continues to be laminar, until at some point, all of a sudden, the flow pattern breaks up and the aforementioned whorls/eddies develop. The flow has become turbulent. Another way of saying this is that the flow moves so fast that viscosity is not able to keep it together any longer.

Exercise 6.1 *Find a video on the internet of the onset of turbulence.*

As we shall discuss further in Chapter 8, and have already discussed in Chapter 4, many aspects of atmospheric flow can be described by dimensionless numbers, that is, numbers without a unit, arrived at using physical reasoning. In the case of turbulence the number we need is the *Reynolds*[2] *number*, defined as:

$$\text{Re} = \frac{UL}{v} \tag{6.1}$$

---

[2] The non-dimensional number was actually proposed by George Gabriel Stokes in 1851, but Reynolds popularised its use many years later.

Exercise 6.2 *Determine the unit of the Reynolds number.*

To explain the quantities used in the definition of the Reynolds number we say that U is a characteristic velocity and L is a characteristic length scale of the flow.

Having studied many flows it has been determined that the onset of turbulence for atmospheric flow starts at Reynolds numbers around 1000, which means that if the Reynolds number of the flow is above this value, the flow is turbulent.

Exercise 6.3 *What would the Reynolds number of typical atmospheric flow be? The kinematic viscosity at 15°C is $1.48 \cdot 10^{-5}$ m^2/s.*

As we have just seen, solving the exercise, atmospheric flow is literally always turbulent, again stressing the fact that understanding turbulence is very important.

Before we start discussing what generates turbulence, I would like to describe a 'trick' that is used when modelling it, and that is called Taylor's hypothesis. The idea is that – for short periods of time – the turbulence field can be considered as frozen, that means that the wind field is moved forwards unchanged (so the time series also becomes a 'space series') (Panofsky and Dutton, 1984).

6.1 What Generates Turbulence?

Turbulence in the atmosphere can be generated through two physical mechanisms:

- Mechanical
- Thermal

The mechanical turbulence is caused by the fact that especially near the surface we have strong shear (as we saw in Chapter 4) caused by the fact that the velocity near the surface is zero and then it increases as we get further and further away – this generates shear in the flow and the flow starts to 'tumble' over, resulting in the mechanically generated turbulence.

The easiest way to describe the generation of thermal turbulence is to look at how the so-called thermals are generated: on a sunny day, different parts of the ground are heated slightly differently, resulting in that the relatively hotter air rises. This process is quite violent, resulting in the generation of turbulent eddies, big as well as small. You sometimes see birds and gliders use these to gain altitude. Thermal turbulence has been generated.

So, we have described two quite different mechanisms producing the same outcome: turbulent flow. In reality it is often a combination of the two mechanisms that typically causes the turbulence.

6.2 Reynolds Decomposition and Averaging

As I mentioned in the introduction, we need to look at turbulence using amongst others the lens of time-series analysis. As an example of what we are looking at, think back at the time-series of wind speed we discussed in Chapter 3. And think, to be more specific, about the plot of the x-component of the wind speed versus time, this is the type of signal we want to look at here.

To ensure that we are only looking at the turbulent part of the flow, we take out the mean (meaning the average over time) of the flow, because whether the mean wind in itself is high or low does not matter, we are only interested in the turbulent part. This way of splitting the time series is called *Reynolds decomposition*. Mathematically this can be written:

$$u = \bar{u} + u' \qquad (6.2)$$

where u is the wind speed, $\bar{u}$ the mean, and u' what remains when the mean wind has been subtracted, that is, the turbulence.[3]

Bio 7: Richardson, Lewis Fry[4]
1881–1953
Newcastle upon Tyne, Northumberland, England
British mathematician, physicist, meteorologist, psychologist and pacifist who was the first to apply mathematical techniques to predicting the weather accurately.

Exercise 6.4   *Look at the time series in Figure 6.2. Try to draw a straight horizontal line through the time series where as much of the time series is above the line as is below. This line will be the mean. Looking at the signal with this line as the reference, we have identified the turbulent part of the signal, using Reynolds decomposition. From now on, when we talk about the time series we mean the u' part only, that is, using the line we have just drawn as the new reference.*

Exercise 6.5   *What is the mean of u'?*

If we now go back to the Navier–Stokes equations in Box 14, and insert this new way of splitting the wind speed into the equation, and then time average the terms (following a specific procedure called Reynolds averaging) we end up with the so-called Reynolds-averaged Navier–Stokes (RANS) equations which were the ones we discussed in Chapter 5.

---

[3] Digging a bit deeper we need to make sure that the time series we are looking at has no trends, that is, is stationary.
[4] Figure source: "Lewis Fry Richardson" by NOAA – NOAA presentation, http://commons.wikimedia.org/wiki/File:Lewis_Fry_Richardson.png#/media/File:Lewis_Fry_Richardson.png [Public domain], from Wikimedia Commons.

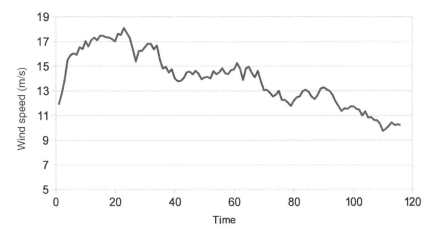

Figure 6.2 A part of a time series

6.3   Spectra

We have now isolated the part of the time series that has been defined as the turbulent part. We will return shortly to the more traditional time series analysis tools, but in this section we will first discuss the strongest tool used in analysing turbulence: the spectrum. I will give one of the usual hand-waving introductions to spectra, and show some standard spectra, too. The anatomy of the different parts of the spectrum will also be explained.

Bio 8: Reynolds, Osborne[5]
1842–1912
Belfast, United Kingdom
Numerous contributions to mainly fluid mechanics, but also other fields. Professor of engineering at Manchester University. Educated from University of Cambridge, Queens' College, Cambridge, Victoria University of Manchester.
Photo of a painting by John Collier.

[5] Figure source: "OsborneReynolds" by John Collier. Original uploader was Seanwal111111 at en.wikipedia - Copied from johnbyrne.fireflyinternet.co.uk. Cropped photo of a painting of Osborne Reynolds painted in 1904 by John Collier. Transferred from en.wikipedia; transfer was stated to be made by User:Stefan Bernd, http://commons.wikimedia.org/wiki/File:OsborneReynolds.jpg#/media/File:OsborneReynolds.jpg [Public domain], from Wikimedia Commons.

If it has been a long time since you have worked with maths, the following might be a bit daunting, but please give it a try. I have done my best to keep things as simple as possible.

6.3.1 Understanding Fourier Analysis and Spectra, a Poor Man/Woman's Approach

You can skip this section, but it will give you a simple introduction to spectra, which are an essential part of understanding turbulence, so please give it a try.

As mentioned above, one of the main ways of describing and analysing turbulence is through the so-called turbulent spectra. The mathematics is quite complicated, but the idea is very simple, so in this section we are going to get a quick idea about how spectra work.

Spectra are a representation of a time series in the so-called frequency domain. If you have forgotten what frequency means, do not worry, I will explain shortly. To get to the frequency domain, we need a mathematical tool called Fourier analysis. The basic question that the Fourier analysis asks is: assume that all time series can be looked at as being made up of a number of waves of different frequencies. Which ones can be found in the time series we are looking at and how dominant are they? This involves a bit of complicated calculations (I will show you the equations in Box 17), but here we will go through a simple example to identify just the frequency part.

Remember that the relation between the frequency, f, measured in Hz and the period, P, measured in seconds, is given by[6]:

$$f = \frac{1}{P} \tag{6.3}$$

Frequency measures how many times the wave goes up and down per second and the period how much time there is between each wave (top).

In the first simple case, our entire time series is just a single sine wave (going from $-\infty$ to ∞), this is shown in Figure 6.3. So the first question is: what is the frequency that lies behind this time series? Please give it a try, I have included a lot of lines in the plot to help you.

As you can see from the figure, there is a top every 20 seconds on the graph, this means that the period is 20 seconds. Getting from period to frequency is just: $f = 1/P$, so the frequency is 0.05 ($= 1/20$) Hz, we have therefore identified the one frequency that makes up the time series. In other words, instead to plotting the whole time series, it can be represented by one number in frequency space. The Fourier analysis would then go one step further and say that the frequency 0.05 Hz has a strength of a certain magnitude.

If we were to plot the so-called spectrum of this simple time series, which is a plot with frequencies along the x-axis and the strength of the particular frequency on the y-axis, it would be very simple in this case as shown in Figure 6.4.

Taking this one step further is to have our time series made up of two sine waves overlaid as shown in Figure 6.5. We now need to ask: which two frequencies can we identify? Again, please give it a try.

---

[6] And if we know the speed, s, we can get to the wavelength, λ, via: $\lambda = s/f$.

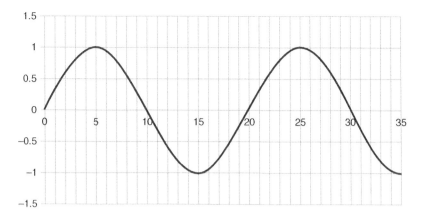

Figure 6.3   A part of the sine wave going from −∞ to ∞

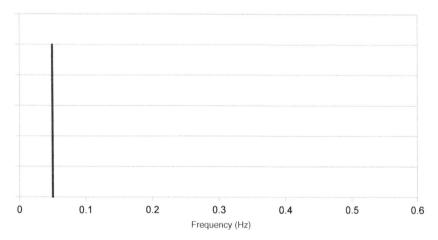

Figure 6.4   The simple spectrum of the sine wave mentioned in the text

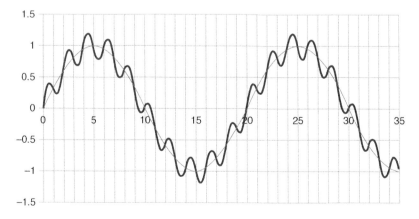

Figure 6.5   A part of the time series made up of two sine waves going from −∞ to ∞

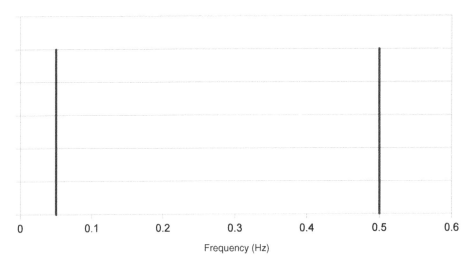

Figure 6.6   The simple spectrum of the two-sine-wave time series

You can probably see that the sine wave from before is still there (so one of the frequencies is again 0.05 Hz), but on top of it, there is another smaller wave. The distance between the tops of the smaller wave is 2 seconds, meaning that the frequency is 0.5 (= 1/2) Hz.

Plotting again our primitive type of spectrum for the time series we would get the spectrum shown in Figure 6.6. That is, the signal is now made up of two frequencies (note that the bar only indicates that the frequency is present, not how dominant the corresponding wave is).

Taking the last step we will look at a time series made up of a series of random numbers as shown in Figure 6.7.

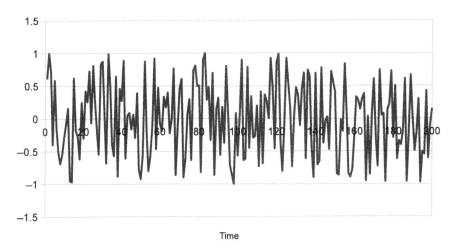

Figure 6.7   A time series made up of random numbers

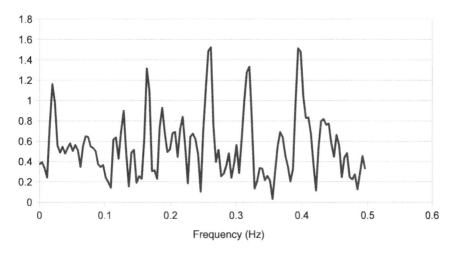

Figure 6.8   The spectrum of the random-data time series

A proper Fourier analysis (using the so-called Fast Fourier Transform, FFT) would result in the spectrum shown in Figure 6.8. This is what is called the proper *spectrum* of the time series, where we can also see the strength of each of the frequencies in the plot.

Box 17   Fourier analysis
{*you can easily skip this box*}

The mathematical expression of the Fourier transform, $\hat{f}(\xi)$, of a function, $f(t)$, is:

$$\hat{f}(\xi) = \int_{-\infty}^{\infty} f(t)\, e^{-2\pi i t \xi}\, dt$$

so a bit more complicated that our simple explanation! Basically this expression does the same as we did, but instead using the e^{ix} function, where i is the complex unit ($= \sqrt{-1}$). The complex exponential can be translated into a sine and a cosine function, connecting us back to our simple spectrum above.

6.3.2   Standard Types of Spectra

We have now spent quite a bit of time understanding the spectrum and the ideas behind and if we look at atmospheric winds various people and organisations have come up with various standard types of spectra representing the turbulent wind.

As we shall see shortly, it is not that easy to measure turbulence, and in particular to measure at all the scales required to make a complete spectrum. Therefore it is very useful for many applications (like load calculations, see later) to have these standard spectra.

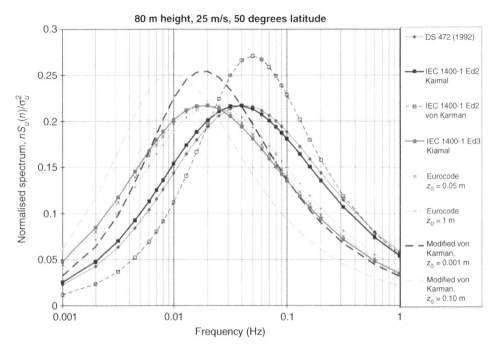

Figure 6.9 Different types of spectra. Source: Burton et al. 2011. Reproduced with permission of John Wiley & Sons, Ltd

An example of a number of such standard spectra can be seen in Figure 6.9. As you can see they have all more or less the same shape, peaking in slightly different locations, and distributing the energy slightly differently. No more on this here, but as always, remember to ask what people mean when they say a turbulent spectrum. Note that the von Karman spectrum is the same person as in the von Karman constant, κ.

This is yet another way of describing Richardson's poem, the way the spectrum is plotted means that it shows what the energy content is for a given frequency, and as you can see we find energy at all scales, just as in the poem.

It is actually surprising and quite remarkable that the spectrum is so smooth (compared e.g. to the spectrum of the random time series we saw above). This means that there is a lot of order in turbulence, just of a different kind than we are used to.

You can also see from the plot that there are distinct parts of the spectrum (c.f. Figure 6.10):

- Energy-containing eddies (integral length scales)
- Inertial subrange eddies, $-5/3$ law (Taylor microscale)
- Dissipation range (Kolmogorov microscale)

These different scales each identify the three parts of the atmospheric spectrum.

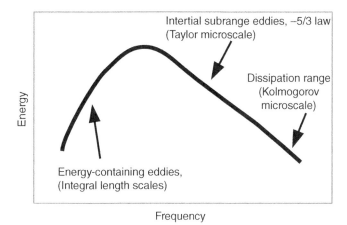

Energy

Intertial subrange eddies, −5/3 law
(Taylor microscale)

Dissipation range
(Kolmogorov
microscale)

Energy-containing eddies,
(Integral length scales)

Frequency

Figure 6.10   The anatomy of a generalised spectrum. The three parts of the spectrum, further described in the text, are identified

---

Box 18   Eddy viscosity
{you can easily skip this box}

There is a term that is very widely used when talking about modelling, especially for the so-called RANS models as mentioned in Section 5.2, and that is *eddy viscosity*.

We know viscosity from the molecular level, and we have met it in the Reynolds number. In popular terms, viscosity describes how 'thick' a fluid feels (honey (high) versus water (low)), a bit more technically it describes the fluid's resistance to being changed by shear (or pulled).

Transferring this picture to turbulent motion, and going to a scale much larger than the molecular, we can call the flow's resistance (note that we are now talking about the flow, rather than the fluid) to transferring momentum via turbulent eddies the eddy viscosity.

The simplest way of describing this mathematically is through the so-called K-theory:

$$\overline{u'w'} = -K\frac{\partial \overline{U}}{\partial z} \qquad (6.4)$$

where K is the eddy viscosity. Typical values for K are 1 m²/s, which as we saw in Exercise 6.3, is 10,000 times bigger than the kinematic viscosity of the atmosphere. The expression says that there is a relation between the fluxes $\overline{u'w'}$ and the shear of the flow. This is of course a simplification, but it works in many cases when the eddies are not too large (as they will be in the thermals we discussed above).

Bio 9: Jean-Baptiste Joseph Fourier[7]
1768–1830
Auxerre, Burgundy, Kingdom of France
French mathematician and physicist, best known for his work with Fourier series. The
Fourier transform and Fourier's law are also named after him. Fourier is also one of the
first, if not the first to discover the greenhouse effect (see Section 4.8.1).

6.4 Measuring Turbulence

As we saw in Chapter 3, different instruments can measure changes in the quantities they
measure with different time resolutions. We have also just learnt that turbulent motion happens
on many different scales, including the very short. This of course means that in order to
capture as many of the scales as possible we need instruments that are able to resolve these.
Unfortunately, our good friend the cup anemometer is not very suited for this purpose (we will
see in a short while what we – any way – can get out of a cup with respect to turbulence), so
we need to look at more 'exotic' instruments.

The most common is the sonic anemometer, which was described in Section 3.4.4. A sonic
can resolve time scales down to 20 Hz, which is a much higher frequency than the cup is able
to resolve (which is typically around 1 Hz).

A good question to ask is: how can you actually measure turbulence, with the many time
scales involved? The spectrum shows, as we have already discussed, that even when we zoom
in a lot on the time series there is still a lot of energy (or motion) present. So if you are familiar
with the so-called fractals, time series of turbulent wind has a fractal dimension higher than
one (Tijera et al., 2012).

Again, the answer is not direct, and instead of measuring all the small and very small
fluctuations, we return to averaged quantities. The most common, and the one that is used in
various standards (see next section) is the *turbulence intensity*, TI, which is defined as

$$\text{TI} = \frac{\sigma_u}{\bar{u}} \tag{6.5}$$

where σ_u is the standard deviation of the u component of the wind speed, and $\bar{u}$ the mean wind.
Generally one would expect that the higher the wind speed the more turbulent the wind. This
of course does not say so much about the turbulence itself, so by dividing/normalising with

---

[7] Figure source: Bunzil, https://commons.wikimedia.org/wiki/File:Fourier2.jpg, at en.wikipedia [Public domain],
from Wikimedia Commons.

the mean wind we try to take some of this effect out. So, the turbulence intensity, is – as the name indicates – a more direct measure of how intense the turbulence is.

Exercise 6.6   *What is the unit of turbulence intensity?*

Box 19   Turbulence on Mars

To give you a bit of a break in the middle of all this talk about turbulence, I am going to tell you a story of how I first met turbulence in a real scientific context, and that was through the study of winds on the planet Mars! As you have seen, there are various models for how the spectrum of turbulence should look. At university we were told that these theories about turbulence in many ways were universal, because they were based on very fundamental relations. Living on a very small planet in a very big universe, this is of course a bit of a sweeping statement! So, when the research lab I worked for (Risø in Denmark) got the chance to get our hands on wind data from the planet Mars, it was very obvious to try to see if our universal theories actually also worked at least on one other planet in the universe.

The data we used was taken from the hot wire anemometer (marked with the circle) from the Viking Lander (NASA, 2015), which we have already met in Chapter 3.

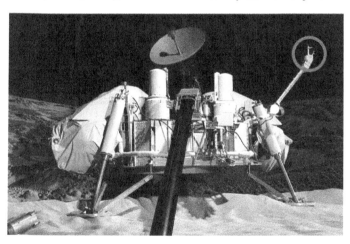

(artist's concept of the Viking lander. Source: NASA. http://www.nasa.gov/sites/default/files/images/585497main_PIA09703_full.jpg)

After having retrieved the data, isolated the appropriate parts, done the modelling and analysis, we came to the conclusion (Tillman et al., 1994) that, yes, our universal theories did in fact also work on Mars.

When you work with the data on a daily basis it is just numbers, of course, but it takes very little thinking to imagine how crazy this is: come up with the idea to send a lander to Mars, go through with the whole project, actually manage to get to Mars, land the lander in one piece with functioning instruments and then get the data back to the Earth. Fantastic to have been part of it!

One of the clever things about σ_u is that most instruments, including the cup anemometer, can be used to estimate this value, based on the measurements it takes. Likewise, researchers are also trying to back out σ_u information from Lidars and Sodars (for Lidars, see Sathe et al., 2014). All this means, is that we can get some idea about the general characteristics of the turbulent part of the flow, even though we are forced to consider it as a bit of a black box. The way the two axes of the spectrum are normally scaled, means that the area under the spectrum is equal to the total standard deviation, which hints at why it is possible to make this relation.

Box 20   A few tricks concerning σ_u and TI

In this box I shall describe a few relations that can be used when looking for rough estimates of some of the quantities we are discussing here. Note that these are just rough estimates, so use them with caution, please.

The first set of relations is concerned with the standard deviation of the velocity components, the σ's, and are as follows:

$$\sigma_u = Au_*$$
$$\sigma_v \approx 0.75\sigma_u$$
$$\sigma_w \approx 0.5\sigma_u$$

where A is around 2, depending on the roughness.

The second relation gives a very simple – and definitely not always correct – expression for the turbulence intensity, TI, and is a function of height and the roughness:

$$TI = \frac{1}{\ln\left(\dfrac{z}{z_0}\right)}$$

Just to reiterate: these expressions are very approximate, so use them with care!

6.5   Turbulent Loads

We are interested in turbulence, first of all because it is a very important part of the study of atmospheric flow, especially in the boundary layer. But there is another, very crucial, second reason for being so interested in it, and that is because turbulent flow, when it meets the wind turbine, is considered a *load*. I will not go into a lot of detail about loads here, but a load on a turbine basically means that it gets worn. The higher the load, the more wear and tear, also, the longer the loads last for, the more wear and tear. Turbulent is so important for the design of wind turbines that it is one of the two main characteristics that are used when describing the design type of a turbine. Turbines are classified according to the so-called IEC standard (IEC, 2005), the basic statistics are shown in Table 6.1.

Table 6.1 IEC classification, IEC 61400-1, edition 3. V_{ref} is the 50-year maximum 10-minute average wind speed. I_{ref} is the expected value of the turbulence intensity at a wind speed of 15 m/s. A given set of V_{ref} and I_{ref} for a turbine location will give the corresponding IEC class

Wind turbine class	I	II	III	S
V_{ref} (m/s)	50.0	42.5	37.5	Values specified by the designer
A, I_{ref}		0.16		
B, I_{ref}		0.14		
C, I_{ref}		0.12		

Source: © 2005 IEC Geneva, Switzerland. www.iec.ch, reproduced with permission of IEC

As you can see, knowing the turbulence at your site is crucial for the type of turbine you will be able to put up. Not only do we need to know the overall turbulence for the wind farm site, but since it can vary a lot over the site, we actually need to know it at each turbine location. In many cases we also need to have the directional variation, since in some sectors the turbulence could be significantly higher than in others.

Exercise 6.7 *Think of some causes for high turbulence in a specific sector.*

I have listed a few more types of loads in Box 21.

Box 21 Other loads on wind turbines

There is a long list of other loads that have an effect on the wind turbine, please consult appropriate sources, but the list includes: wind shear, directional veer (including jets), wind speed in itself, density, flow inclination, turbulence (as we have discussed here), the way the turbine is operated, and many more.

6.6 Extreme Winds

Extreme winds do not have that much with the physics of turbulence to do. However, the effect on the turbines (loads) is in the same category, so in this section I am going to take you through the most basic parts of the subject.

An extreme wind is some high wind that occurs rarely. In more theoretical language, for example, the fifty-year extreme wind is the maximum encountered wind in a 50-year period. These winds are caused by severe storms, hurricanes/typhoons and other strong weather phenomena. Relating this back again to the wind turbine, the effect on the turbine is quite different from the loads generated by turbulence, the extreme winds try to 'break' the structure, whereas the turbulent loads try to 'exhaust' the turbine.

It requires a very complex statistical procedure to calculate the extreme wind of a site, and again I will not go into any details here. If you want to have a first coarse guess you can use the following formula:

$$u_{50} = 5\bar{u} \tag{6.6}$$

where u_{50} is the 50-year extreme wind (i.e. the maximum wind over the 50-year period) and $\bar{u}$ is the mean wind at the site. Warning: use only this formula with great care, it can be very wrong, especially in areas outside the mid-latitudes.

If you want to know more about loads, turbulent as well as extreme-wind-generated, there are a lot of good books on the topic, for instance Burton et al. (2011).

6.7 Summary

Turbulence, the topic of this chapter, is probably the most difficult of this book's topics to understand. And saying that anyone really understands it, is probably a bit of a stretch, if we by understanding mean a capability to do precise calculations and modelling. It was with this in mind that we started quite poetically by describing the cascade of energy in turbulent motion from the very large to the very small scales, we saw this in Richardson's poem, but we also saw it in the spectra that described the various scales concerned with turbulence. We discussed how to measure turbulence, mainly in a roundabout way, by using the standard deviation and the turbulence intensity, and we even took a trip to Mars and back. A few approximate expressions were also introduced, to help you get a feel for the magnitudes of the various quantities. For me, the most important takeaway message about turbulence is that we don't understand it directly, but we have very sophisticated ways of understanding it in a lot of indirect ways.

The chapter finished off with a bit of an orphaned section on loads where we saw two load-generating phenomena: turbulence which contributed to the wear and tear of the turbine, and extreme winds which would try to break the structure.

Exercises

6.8 *Calculate the turbulence intensity, TI, for the following cases:*
$\bar{u} = 10$ m/s; $\sigma_u = 2$ m/s
$\bar{u} = 10$ m/s; $u_* = 0.3$ m/s
$z = 60$ m; $z_0 = 0.1$ m

6.9 *Calculate the extreme wind for the following two cases: a mean wind of 8 m/s and one of 10 m/s.*

6.10 *Estimate the IEC class for a wind turbine located in one of the following three situations:*
$V_{ref} = 41$ m/s, TI = 0.15
$V_{ref} = 50$ m/s, TI = 0.13
V_{ref} unknown, but u = 9.5, TI = 0.1

7

Wakes

It is not possible to start a chapter about wakes without showing the fantastic picture in Figure 7.1, you see it in quite a few places; however, it is so nice, that I thought I would show it here too. The picture shows the Horns Rev offshore wind farm and the wakes associated with many of the turbines, especially the ones in the front row. As we go through this chapter we will, step-by-step, build our understanding, so that we, at the end, will be able to understand what we see (at least in our usual hand-waving way!).

In this chapter, we will start with the most basic first: turbine-to-turbine wakes, or simply the wake behind a single turbine. Here we will cover the three different basic theories that are currently used in the field. After that, we will look at wind farms, that is, collections of more than a few wind turbines. Here things get a bit more complicated, but I shall describe the two most important modelling tools for this as well. By having gone through the first part, you will see that you will be very well equipped to understand how these two models work. At the end of the chapter, we will cover a number of advanced topics, including very large offshore wind farms, how to deal with measurements and related data issues and the effect of stability and terrain on the wakes, and advanced modelling in general. Finally, I will look at how one wind farm interacts with another (called inter-wind farm wakes).

The definition of a wake is that it is the area behind the wind turbine affected by the fact that the wind turbine takes momentum out of the wind, (see Figure 7.2 for a simple illustration and Figure 7.3 for a real example), or in simple terms where the wind is reduced. The topic of wakes is a slightly different topic to all the others in this book. This is because the wakes are caused by (man-made) wind turbines and not, as is the case for all the other topics, by nature itself. All the other topics are wind-turbine independent, but not this one. However, since you are very likely to run into wakes and for the sake of completeness, I have found it essential to have a chapter describing the basics of wakes and how to model them.

Meteorology for Wind Energy: An Introduction, First Edition. Lars Landberg.
© 2016 John Wiley & Sons, Ltd. Published 2016 by John Wiley & Sons, Ltd.

Figure 7.1 The wakes of the Horns Rev 1 offshore wind farm on 12 February 2008. Made visible by foggy conditions at the site and low winds (around 4 m/s) (see Hasager et al., 2013 for a detailed analysis). Source: Vattenfall AB, under a Creative Commons licence. Photograph by Christian Steiness

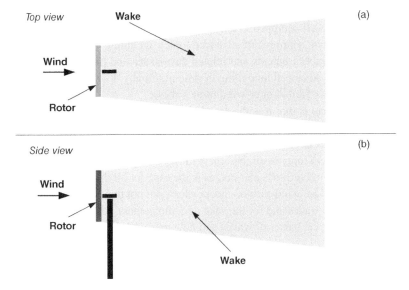

Figure 7.2 Illustration of a simple wake behind a wind turbine, seen from the top (a) and from the side (b). The wake is the shaded area behind the turbine

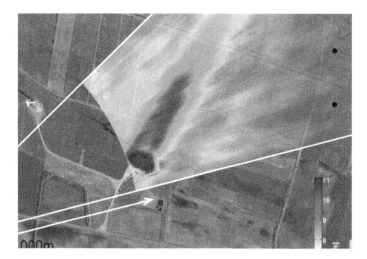

Figure 7.3 The wind speed field behind a single wind turbine scanned with a laser. The darker the colour, the lower the wind speed. The very dark area behind the wind turbine (located in the lower left of the scan) is the wake. Source: Hauke et al. (2014). Reproduced by permission of Forwind, University of Oldenburg

The topic of wakes – especially of very large (typically offshore) wind farms – is probably (in very close competition with more general flow modelling) the topic where currently most research takes place, so expect lots of changes, updates, and new discoveries in this field.

7.1 Turbine-to-Turbine Wakes

In the simplest case, think about two wind turbines, one in front of the other (we actually only use the second turbine as a kind of a measuring device, we could also just have used an anemometer of some sort, mounted on a mast). When the wind blows along the line connecting the two turbines, the one behind the other will be affected by the fact that the front turbine will have converted some of the kinetic energy (i.e. the energy from the moving/flowing motion of the air) into mechanical energy (which in turn is converted into electric energy, the whole purpose of the wind turbine) and thereby reducing the energy available to the second turbine (just to underline this point, note that the wind turbine *converts* the kinetic energy into mechanical energy, i.e. no energy is *produced* as such). As you get further and further away from the front turbine, more and more kinetic energy/momentum is mixed in from the surroundings, and the reduction decreases gradually (through the turbulent motion of the atmosphere, see Chapter 6). This is the basic idea of wakes and the theory behind wake modelling; refer back to Figure 7.2 for a schematic illustration and Figure 7.3 for a laser scan of a real single wake.

If you have ever seen a wind turbine in action you can probably imagine that, right behind the wind turbine, the flow is very complicated (this is right after the wind has left the rotor area, and has gone through a dramatic encounter with the blades of the wind turbine). Because of this and because of the simplifications made in developing the models, most models are not valid right behind the rotor. Normally, this area goes out to the first two to four rotor diameters downstream. In general, when talking about wakes it is very common to measure most of the distances in rotor diameters, that is, the diameter (approximately two times the length of the blade) of the rotor. This means of course that this area right behind the wind turbine is a grey area, unknown to modelling; however, most wind farms would never see a spacing of the wind turbines so close (there are a few close-spacing exceptions), so the practical consequences of this assumption are limited.

Digging a bit deeper we find three different approaches to the modelling of wakes.[1] The first one, known as the NO Jensen model[2] is quite simple and based on the so-called momentum theory (Jensen, 1983), the other – a bit more advanced – is the theory by JF Ainslie[3] (Ainslie, 1988), based on the eddy viscosity closure (which is a simplified version of a full CFD model as we discussed in Chapter 5, see Box 18). Finally, a model based on a more generic type of theory, called similarity theory (which we also met in Chapter 4 when discussing stability), will be described. This last one is based on first principles, but gives a good feeling for how a wake (or rather a model of a wake) should behave, and it should probably be considered more a hand-waving kind of theory, informing how models could be built, rather than a model proper.

As you will see in the following, all the models describe how the wake (mainly the reduction in the velocity right after the turbine and the gradual increase thereafter and the expansion of the width of the wake) develops as you get further and further away (downwind) from the wind turbine.

In the following, the three models will be described in a bit more detail. However, there will be a lot of equations, so I will try to explain what these equations describe, and we will also do some exercises along the way, to see a bit more about the models and how they work and differ.

7.1.1   The NO Jensen Model

We will start with the model by Jensen (the one that was called the NO Jensen model, which is also known as the PARK model), which says:

$$u_w = u_i \left[1 - (1 - \sqrt{1 - C_t}) \left(\frac{D}{D + 2kx} \right)^2 \right] \qquad (7.1)$$

The expression in Equation 7.1, has one well-known variable, u (the wind speed), and a few we have not seen before. But first the u's: u_w is the wind speed in the wake at downwind distance

---

[1] There are of course many, many more approaches, but the ones I have chosen cover the different types of models quite well.

[2] Named after Niels Otto Jensen of Risø (now DTU Wind) in Denmark.

[3] Named after JF Ainslie of the Central Electricity Generating Board (now broken up into National Grid Company and a few other companies) in England.

x, and u_i, the speed of the wind right before (upwind of) the turbine (this is also called the free-stream wind speed). So, the equation tells us how the upstream wind speed is reduced due to the wake of the turbine. D and C_t have to do with the physical/mechanical characteristics of the wind turbine in question: D is the diameter of the rotor and C_t the *thrust coefficient* (see Box 22). The final variable k (called the wake decay constant, see later in this section), is a parameter that actually depends on a lot of things, but in this simple description, the standard definition will be used:

$$k = \frac{A}{ln\left(\frac{h}{z_0}\right)} \tag{7.2}$$

which basically says that the parameter k depends on the roughness, z_0 (see Section 5.3), and h, the height of the wind turbine. A is a constant ($= 0.5$).

Box 22 C_p and C_t

There are two coefficients that are used to describe the characteristics of a wind turbine:

C_p, the power coefficient (rotor power performance), which is equal to the power the rotor generates (P) divided by the available power in the wind:

$$C_p = \frac{P}{\frac{1}{2}\rho u^3 A}$$

C_t, the thrust coefficient, which is equal to the thrust force (T, basically the difference between the momentum before and after the rotor disk) divided by the dynamic force. (Note that the wind speed is squared here, as opposed to cubed above):

$$C_t = \frac{T}{\frac{1}{2}\rho u^2 A}$$

So, to hand-wave a bit: the wake downstream of a wind turbine depends on the wind speed before the turbine, the distance (inversely proportional to the square of the distance), the wind turbine itself and the type of terrain the wind turbine is located in – which I guess all makes a lot of sense.

Exercise 7.1   *As an example, let us consider a 100 m high turbine, with 50 m blades and a C_t of 0.7. Let us assume that the roughness is 0.03 m (grass). Plot the wind speed in the wake as we go downstream of the turbine from 2D to 500 m (5D). The wind speed right before the turbine is 10 m/s.*

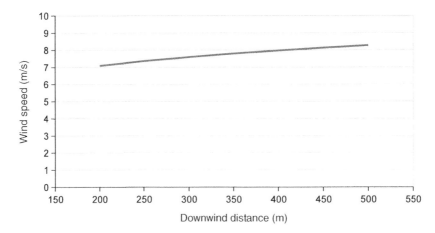

Figure 7.4 The behaviour of the wind speed in the wake of a turbine, calculated using the NO Jensen model (Equation 7.1), see Exercise 7.1 for further details

We are going to do the exercise here (but as always, please give it a try yourself first): k can be found from

$$k = \frac{0.5}{ln\left(\frac{100}{0.03}\right)} = 0.062 \tag{7.3}$$

Putting Equation 7.1 into a spreadsheet, inserting the values, and plotting the results, we get the plot shown in Figure 7.4.

Maybe not the most interesting plot, but as we expected, the wind speed in the wake is much reduced right behind the turbine (remember we should not trust the model closer than a minimum of 2D to the turbine), and as we get further and further away, the wind picks up again. Note the size of the reduction, a wind speed of about 7 m/s, means a 30% reduction of the free wind speed. Note also that at 5D, we are still not back up at the original wind speed of 10 m/s.

Returning to the wake decay constant, k, it is also used in the description of how the wake widens as we get further and further away from the turbine. As mentioned above, the NO Jensen theory is quite simple (but works very well in many cases), so it assumes that the wake expands linearly, that is, following a straight line, with the width, W, growing as:

$$W = D + 2kx \tag{7.4}$$

Exercise 7.2   *Plot the width of the wake as a function of distance, using the NO Jensen model. Assume the same values as above.*

Since we will need the numbers a bit later, we will do the exercise here: inserting the values from above, we get the plot shown in Figure 7.5.

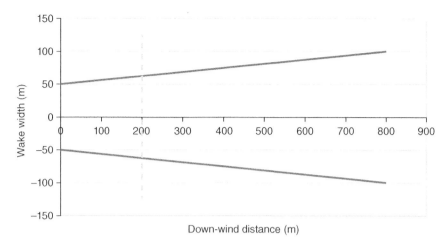

Figure 7.5 The expansion of a wake behind a wind turbine using the NO Jensen theory (Equation 7.4), see Exercise 7.2 for more details

As you can see, and probably have expected, the width of the wake increases, as we get further and further downstream from the turbine, due to the turbulent mixing in of air with momentum unaffected by the turbine.

7.1.2 The Ainslie Model

The second of the three wake models is the Ainslie model, and it also describes the development of the wake as we go downstream; however, the model is quite a bit more complicated, using more advanced fluid dynamical equations (see Tennekes and Lumley (1972) and WindFarmer theory manual (2015),[4] for the full story), so we are only going to look at the model at one particular place, at 2D (i.e. right where the model starts to be valid), which is where an empirical formula exists for the velocity deficit (i.e. how much has the wind been reduced compared to the wind right in front of the turbine), the formula says:

$$D_m^i \equiv \frac{u_i - u_w}{u_i} = C_t - 0.05 - \left((16C_t - 0.5)\frac{I_0}{1000} \right) \qquad (7.5)$$

where D_m^i is the initial centreline velocity deficit (same u's as before), again C_t is the thrust coefficient, and I_0 is the ambient turbulence intensity in percent (a normalised measure of how turbulent the wind is; see Chapter 6 for more on this). Going further downstream, the more advanced CFD model is used, and the expression above together with the shape of the profile, described below, are used to initialise (i.e. start) the model.

---

[4] Disclosure: I currently work for the company that develops and sells this model.

Exercise 7.3   *Compare the wind speed at 2D downstream using Equation 7.5 to what we found in Exercise 7.1 above. Assume same wind speeds, C_t as above, and $I_0 = 10\%$.*

Again we will do the exercise here (but please try yourself first): inserting the values in the equation we get

$$D_m^i = 0.7 - 0.05 - \left((16 \cdot 0.7 - 0.5)\frac{10}{1000}\right) = 0.54 \tag{7.6}$$

and calculating the wind speed based on this we get

$$u_w = \left(1 - D_m^i\right)u_i = (1 - 0.54) \cdot 10 = 4.6 \text{ m/s} \tag{7.7}$$

Comparing this to the 7 m/s we found above, we can first see that in both cases, there is a significant reduction right behind the wake, but we can also see that the two models give quite different results. You will see shortly that the two models assume rather different shapes of the wakes, and that could be part of the explanation. Note, that in both cases, we are very close to the turbine, and the validity of the models is just starting to kick in.

As opposed to the NO Jensen model, the Ainslie model makes the more 'natural' assumption that the shape of the wake is Gaussian (as opposed to the straight line of the NO Jensen model). The width of this wake, b, is given by

$$b = \sqrt{\frac{3.56C_t}{8D_m(1 - 0.5D_m)}R_r} \tag{7.8}$$

where D_m is the centerline velocity deficit at any point and R_r the radius of the rotor. The expression for the profile is given by

$$p_w = 1 - D_m \exp\left(-\left(\sqrt{3.65}r/b\right)^2\right) \tag{7.9}$$

Since this is getting a bit complicated again, let us do another exercise:

Exercise 7.4   *Using the wind speeds, etc. from Exercise 7.3, draw the profile of the wake deficit (i.e. the profile that is found perpendicular to the wind direction) at 2D.*

Doing this right away again, and inserting the values in the expression for b, we get:

$$b = \sqrt{\frac{3.56 \cdot 0.7}{8 \cdot 0.54(1 - 0.5 \cdot 0.54)} \cdot 50} = 89.9 \tag{7.10}$$

and inserting this in the expression of the profile (Equation 7.9) we get the profile shown in Figure 7.6.

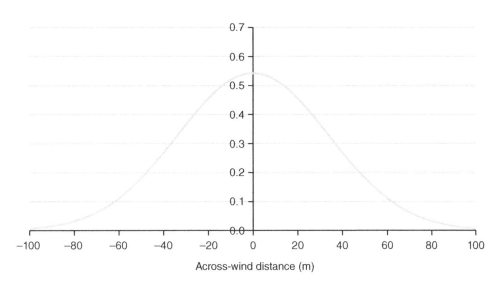

Figure 7.6 The profile of the wake deficit at 2D downwind of a wind turbine using the Ainslie model (Equation 7.9)

It might seem a bit counter-intuitive that the maximum is in the middle, but remember the 'shape of the wake' is the function that describes the wake *decay*, so the wind speed profile will look differently, opposite in some sense. We will now see an example of how the actual wind speed profile looks.

Exercise 7.5 *Using the information above, draw the wind speed profile at 2D again using Ainslie's model, and compare to the wind speed profile of the Jensen model.*

Given the deficit, the velocity in the wake can be found from (manipulating the first part of Equation 7.5):

$$D = 1 - \frac{u_w}{u_i} \Rightarrow u_w = u_i(1 - D) \tag{7.11}$$

which, when plotted and using the same numbers as above, gives the plot in Figure 7.7.

Note that the wind speed profile of the Jensen model is just a straight line.

So far, we have looked at the wakes in quite a two-dimensional way, going either along (downwind of the wind turbine) the wake, or across. Since all this takes place in a three-dimensional world and the wind turbine is a three-dimensional structure, the wake of course is three-dimensional as well. In order to get a feeling of what the three-dimensional structure of the wakes of the two models look like, I have plotted them in their full three-dimensional form in Figure 7.8.

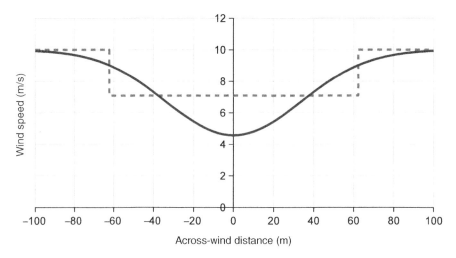

Figure 7.7 A comparison of the Ainslie (solid line) and the NO Jensen (dashed line) model for the wind speed profile in the wake of a wind turbine 2D downwind

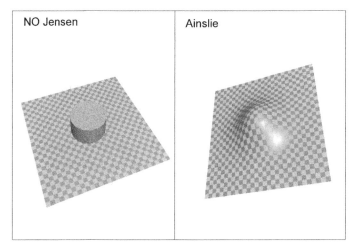

Figure 7.8 Illustration of the two-dimensional velocity deficit for the two models: NO Jensen (left) and the Ainslie model (right). The velocity deficits are shown in a plane perpendicular to (i.e. crossing) the wake. The plots are not to scale and are just intended to illustrate the differences in the shape of the fields calculated by the two models

7.1.3 Similarity Theory

Similarity theory is a very general type of theory in physics; it is based on, surprise, surprise, the fact that many quantities are similar for many systems. A very simple – often used – example is from geometry where similar shapes, e.g. triangles, will have similar *relations* between various aspects of the shape (e.g. the ratio of the sides between a small and big triangle of similar shape). Another aspect of similarity theory is to look at *dimensions* of different quantities, again a simple example could be that a speed is made up of a typical length scale relative to a typical time scale; reasoning like this is often used in fluid dynamics. Finally, these theories also often come up with dimensionless numbers, like the Reynolds number which we discussed in Chapter 6, which can be used to describe these similarities in a non-dimensional way. The nice thing about similarity theories is that they often come up with scaling laws and simple exponential functions that express the behaviours of different quantities.

I do not wish to go into much more detail here, but if you want to know more, you can go all the way back to the roots of similarity theory used in boundary-layer meteorology and consult Monin and Obukhov (1954).

Using similarity theory on wakes (see Abramovich, 1963 or Pope, 2000) we get the following theoretical expressions as a function of downstream distance.

The width of the wake, w, will expand with downwind distance, x, as

$$w \propto x^{1/3} \tag{7.12}$$

the wake deficit, D, decays as

$$D \propto x^{-2/3} \tag{7.13}$$

and finally, the wake profile, p, perpendicular to the wind, as a function of the radius, r, will look as follows

$$p \propto \exp\left(-r^2 \ln 2\right) \tag{7.14}$$

Note that we are now using the *proportional to*, $\propto$, symbol rather than the $=$ one, to underline that these relations describe the qualitative behaviour rather than the quantitative ditto.

To see how this compares to the Jensen model we will do three quick exercises:

Exercise 7.6   *Plot the width of the similarity profile versus the NO Jensen model.*

I will do it here, but please also give it a try yourself first. Plotting the numbers we get the plot shown in Figure 7.9.

The dashed grey line is the result of the Jensen model (from Exercise 7.2) and the solid black line is the similarity theory prediction. I have scaled Equation 7.12 in various ways to make it fit the Jensen model.

Exercise 7.7   *Plot the deficit as a function of the down-wind distance, and compare with the NO Jensen profile.*

using the values from Exercise 7.1, and the expression in Equation 7.13 we get the plot in Figure 7.10. Again I have scaled Equation 7.13 so as to match the NO Jensen model.

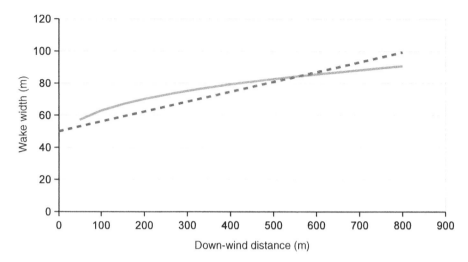

Figure 7.9   The width of the wake as we go downwind calculated using the NO Jensen model (dashed line) and similarity theory (solid line). The similarity model is scaled to fit the NO Jensen model

Exercise 7.8   *Finally, compare the across-wind profile of the similarity theory with that of the NO Jensen profile.*

Using the numbers from Exercise 7.1 and Equation 7.14, we get what is shown in Figure 7.11. As above I have scaled the equation to match the NO Jensen model.

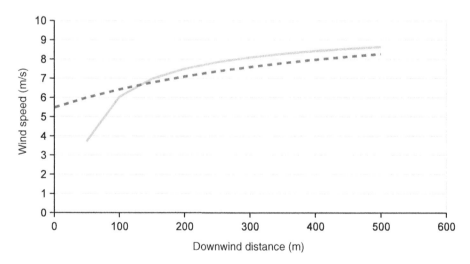

Figure 7.10   The wind speed downwind of a wind turbine, calculated using the NO Jensen model (dashed line) and similarity theory (solid line). The similarity model is scaled to fit the NO Jensen model. The similarity theory is probably the more correct of the two at the beginning of the wake

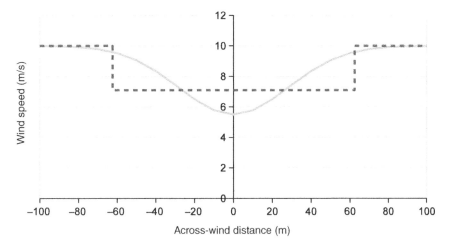

Figure 7.11 The across-wind profile of the wake, using similarity theory (solid line) and the NO Jensen model (dashed line)

7.1.4 Effect on Power

Up to this point we have looked at the wind speeds only, and how the wake reduces these, however, what the point of the wake calculation really is, is of course to be able to calculate the reduction of the *power* produced, this is called the *wake loss*. To make that connection we need to introduce the wind turbine *power curve*, which is the curve that relates the wind speed at hub height to the power produced (I know, I should have said converted) by the wind turbine given that speed. A generic example of a power curve can be seen in Figure 7.12.

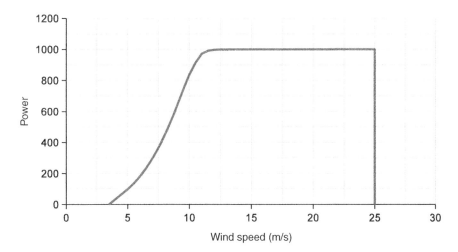

Figure 7.12 A generic power curve. Wind speed at hub height along the *x*-axis, and the power produced along the *y*-axis

Table 7.1 Consequences of reducing the wind speed by 15%. Column 1 is the wind speed at hub height, column 2 the corresponding power, column 3 the reduced wind speed (a 15% reduction), column 4 the reduced power, and the last column gives the resulting reduction in power

Speed	Power	Reduced wind	Reduced power	Reduction (%)
5	98	4.3	40	58
10	836	8.5	538	37
20	1000	17.0	1000	0

If you have never seen a power curve before it is worth to note a few things (otherwise you can easily skip this paragraph): first, the turbine does not start producing power until the wind reaches a certain wind speed (called the *cut-in wind speed*). As the wind speed increases, so does the power, but note that it is proportional to approximately the square of the wind speed (compare this to the derivation we made in Chapter 1, where we found the relation between wind speed and the kinetic energy to be cubic). What this of course means, when discussing wakes, is that the wake loss in wind speed terms is even more pronounced in power terms. The production then levels out (we have reached the *rated power*), and finally, the turbine stops producing (at the *cut-out wind speed*). This is done to save the turbine from excessive loads during very high winds.

Exercise 7.9 *Looking at the power curve in Figure 7.12, estimate what a 15% reduction in wind speed due to for example, wakes would mean in power terms, if the wind speed was 5 m/s, 10 m/s and 20 m/s, respectively.*

(Please pause and do the calculations) Interpolating from Figure 7.12 we get the numbers found in Table 7.1.

So, as we expected the 15% reduction in wind speed could result in significantly higher percentages in power terms (up to almost 60%, and around 40% on the steep part of the curve), but note also, that at the flat part of the curve, a reduction of the wind speed has no effect on the power produced.

7.1.5   Wake Models, Summary

We have looked at quite a few models up until now and seen numerous equations, so to make sure you have captured the main points, I have given a quick summary in Table 7.2.

As you can see – in many ways – the three models are quite similar, with small variations in complexity. To increase this complexity, and, hopefully, thereby get more accurate results, CFD models will need to be used. Currently (in 2015) this is where research is, where the turbines are represented by either actuator disc or actuator line models (c.f. Figure 7.13). In the actuator disc model case, the area swept by the wind turbine rotor is represented by a porous disc (where the thrust is evenly distributed over the swept area, using the thrust-coefficient curve, c.f. Box 22), and in the case of the actuator line model, each of the (three) blades is represented by a line along which the forces are distributed radially. In other words, it is not

Table 7.2 Overview of the three wakes models described in the text

	NO Jensen	Ainslie	Similarity
Describes	– Speed	– Speed	– Speed
	– Width	– Width	– Width
	– Profile	– Profile	– Profile
Wake shape	Line	Gaussian	Gaussian
Complexity	Low	Middle	Low
Valid from	2–4D	2D	N/A

possible to model the effect of the rotor or the blades directly, so these two approximations are used.

A wide range of CFD models are also used, including the very advanced direct numerical simulation (DNS) and large eddy simulation (LES) models (c.f. Figure 7.14). Mesoscale modelling has also been used on some offshore wind farms (Volker et al., 2012).

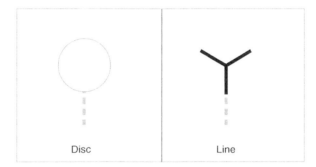

Disc   Line

Figure 7.13 Simple illustration of an actuator disc (left panel) and a line (right panel). The tower of the wind turbine is indicated by the dashed line, since in the model domain, it is only the disc/line that represents the turbine, the tower is ignored

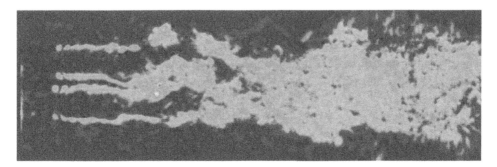

Figure 7.14 Model output from a model that combines large eddy simulations (LES) with an actuator line technique, turbulent inflow.[5]

[5] Figure source: Troldborg et al. 2007. Reproduced with permission of IOP (http://iopscience.iop.org/page/STACKS#metadata) made available under a CC BY 3.0 licence (http://creativecommons.org/licenses/by/3.0/).

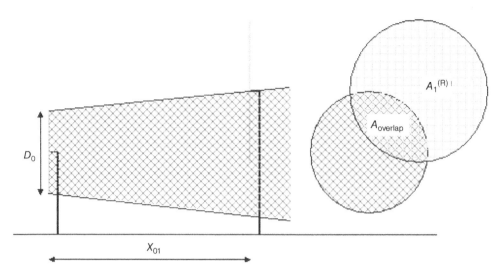

Figure 7.15 Wake addition and overlap as done in the WAsP programme. Source: Mortensen et al. 2014. Reproduced by permission of Department of Wind Energy, Technical University of Denmark

7.2   Several Wind Turbines, that is, a Wind Farm

We have now gone through the basics of wind-turbine-wake theory, and the next natural step to take is to go from the two-turbine case to the multi-wind-turbine case, that is, a wind farm. Here the biggest complications are the fact that we will now see *partially overlapping* wakes and a need to *add* multiple wakes as well, i.e. when one turbine, seen in the direction of the wind, has more than one turbine in front of it (i.e. upstream).

These two complications are dealt with in different ways depending on the model used, and here I will only discuss two: one model, WAsP, which implements the NO Jensen model (Mortensen et al., 2014), and one using the Ainslie model, WindFarmer (WindFarmer, 2015).

In WAsP, this is done simply by adding the fraction of the wakes that overlap the rotor disk in question (Figure 7.15). WindFarmer uses the stronger of the two deficits in the overlapping area.

Because we have done our homework so well, looking at the simple wake theories in great detail, we do not need to discuss the field of wakes in wind farms further, we now have a basic understanding of how this works.

7.3   Advanced Topics

In this last part of the chapter, we will briefly take a look at some of the more advanced topics concerning wakes. The first topic is how we measure wakes, then we look at differences (if any) between off- and onshore wakes, we will also look at very large wind farms, and finally we will look at how one wind farm will affect another. As I mentioned in the introduction, the study of wakes is very much a field where new discoveries, more data and challenges are seen very often, so keep an eye on the literature as well.

7.3.1   Measuring the Wakes

As we will also discuss in the chapter on modelling (Chapter 8), in order to validate – and ultimately improve – any model, observations of the quantities we want to model are required. This is also the case when modelling of wakes is concerned. However, we run into three problems (note, that what we are concerned with here, is mainly the wakes of very large wind farms, since the current models cope quite well with the smaller and thereby simpler wind farms). First, there are not that many large (offshore) wind farms (at the time of writing there are not more than a few handfuls of operational large offshore wind farms in the world), and second, it is really difficult to measure the wakes in the wind farms. Finally, as we have seen, a wake is the reduction of a wind speed, so we are looking at differences between wind speeds rather than absolute values, increasing the uncertainty.

There is not much to do about the first point (other than wait!), however, looking a bit more at the second, currently, what is mainly used are the wind turbines themselves and meteorological masts located in or near the wind farms. As you probably can imagine there are lots of limitations in doing this. Using the wind turbine itself as a measuring device is of course a very good idea, they are there at the wind farm; however, what we measure from a wind turbine is the power it produces and not the wind speed as such.[6] This means that the power needs to be converted back to wind speed (since the models need to predict this more basic quantity), and – looking back at the power curve in Figure 7.12 – you can see that for some values of the power there can be many wind speeds (e.g. below cut-in and between rated and the cut-out wind speed). This can of course be done, but it introduces uncertainties that we do not really want, especially when validating models.

Another, more technical point, is that it is also very important to choose the right width of the directional sector that the data is sampled in, and similarly the right size of the wind speed bin, and the right period of time to average over. The reason for this is that there is a very strong dependence of the behaviour of the wake on these, imagine for example, the difference between the flow exactly along a row of turbines and just a bit off the row. As I said, this is quite technical, but doing this wrongly can lead to very inaccurate conclusions about a model's validity. So, when you look at studies validating a specific model, make sure you understand how direction and speed have been binned and averaged.

Returning to the actual measurement of the wakes, we have already seen that the power output and the wind speed from the turbines could be used, however, not so accurately. Other more fancy instruments include satellite data (for offshore wind farms, see e.g. Hasager, 2014), scanning Lidars and Radars (see Section 3.4.9) and drones of various types (see Figure 7.16 for an example of a drone in a wind farm).

---

[6] There is an anemometer on the wind turbine as well, the nacelle anemometer, and that is also sometimes used. However, the accuracy of these devices and the fact that they are affected by the proximity to the nacelle make them often not good enough to be used for the validation of models.

Figure 7.16 Drone in wind farm. This particular drone does not measure the wind speeds. The Cyberhawk. Source: © Cyberhawk Innovations Ltd. Reproduced by permission of Cyberhawk Innovations Ltd

7.3.2 Onshore/Offshore Wakes

In some physical sense there is of course no real difference between on- and offshore. Offshore is just a flat, very smooth surface compared to most onshore surfaces. However, this is simplifying things a bit too much, mainly due to the fact that there is a very strong tendency to have very large wind farms offshore, and with that the difficulties of modelling very large wind farms are mainly a challenge offshore. The main physical issue offshore is the effect of the thermal properties of water, and the resulting effect on the atmospheric stability (see Section 4.8) which in turn affects the wakes quite significantly. In a nutshell, the problem is that due to the fact that most wind farms have been onshore (where stability is not as important to take into account in most cases), most of the current models have been developed without taking stability into account; in fact we always try to avoid stability-dependent models, to simplify things. This is now being changed, and it is seen that taking stability into account for offshore wind farms really do make a difference (see Barthelmie et al., 2011).

Another effect that has more to do with size than with the fact that the wind farm is offshore, is that the wakes actually *meander*, that is, the wakes do not move in a straight line downwind from the turbine, but swing back and forth. Normally, this is averaged out quite nicely, but the sheer size of the wind farms again means that this effect cannot be ignored. Recently, models to take this effect into account have been developed (see Ott et al., 2011).

On the onshore side of things, a physical feature that is rarely taken into account is the wake/terrain interaction, meaning that the terrain itself changes the shape of the wake, just like the stream lines are affected in general by the terrain (see Section 5.2). Most models would consider the wakes independently of the terrain and that is of course not correct, and in complex terrain cases, probably quite wrong sometimes. Looking back at the section on stability (Section 4.8) you can probably also imagine that stability plays quite a significant role, and hand-waving-wise one can say that in the unstable case, the wakes are being mixed up more quickly, and in the stable case more slowly, meaning as an example that the wakes can persist very far downstream under very stable conditions. Similarly, turbulence will also change how quickly the wake is being mixed up with ambient air: (relatively) high turbulence, a more quick return to upwind conditions, and vice versa for low turbulence.

At the time of writing there is also a discussion about whether there is a 'deep array effect' of these very large wind farms or not, some data show it (Nygaard, 2014), others doubt it. Make sure you keep an eye on the literature to find out the latest.

7.3.3   Very Large Wind Farms, State of the Art

As mentioned in the sections above, the current state-of-the-art is looking at trying to model very large wind farms. In meteorological terms what the issue is, is that when a wind farm gets to a certain size, the effect of it can no longer be considered 'local', it now affects the entire boundary layer (see Section 2.2). This in turn affects the way the models need to be put together.

7.3.4   Wind Farm to Wind Farm Interaction

The last thing I would like to briefly mention in this last section on wakes, is how the wakes behave as they leave the wind farm. This is of specific interest if you are putting up a wind farm near another wind farm (which is often the case offshore, due to planning, logistical and grid connection/substation issues). As you can imagine since there are no turbines left, so to speak, we are looking at wakes which gradually are dying out; the problem, however, is how they die out (and the wind speed consequently picks up). Again different theories can be used (including the similarity theory discussed above), but here I will just show a very simple model for the behaviour of the wake as it leaves the wind farm area: a simple decrease with distance from the wind farm of the deficit (c.f. Figure 7.17). This effect is called intra-wind-farm wakes.

7.4   Summary

As mentioned in the introduction to this chapter, we have moved quite a bit away from the basic boundary-layer meteorological subjects, but as also mentioned in the beginning, I have found it necessary to include this in this book for completeness and because it is a very important issue for estimating the power output of wind farms, which is under intense study currently.

In this chapter we have looked at three models in some detail (the NO Jensen, the Ainslie and the similarity theory models), done quite a few calculations using them, and I hope that I have also given you a feel of how certain (or rather uncertain) these models can be.

Computer codes using these models were also briefly described, mainly to illustrate how multiple wakes interact.

Returning to the picture at the beginning of this chapter we can now understand what we see in much more detail: we can see the wakes developing, we can see how they widen as we go downstream and we can also see (a little bit of) how two wakes interact when they meet and blend. It is quite hard, but maybe not impossible, also to see how the wakes at the far end of the wind farm are completely mixed into each other as they leave the edge of the wind farm.

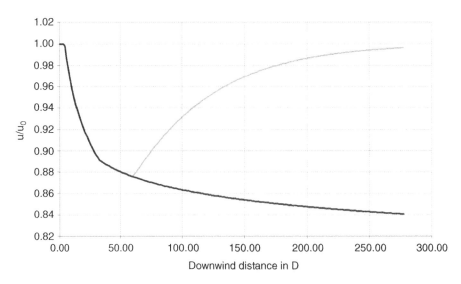

Figure 7.17 The exponential recovery to ambient wind speed of the wind farm wake for a large wind farm as modelled by WindFarmer. Downwind distance along the *x*-axis, wind speed normalised by the upstream wind speed along the *y*-axis. The grey line is the recovering wake. Source: © DNV GL. Reproduced with permission of DNV GL

Exercises

7.10 *Going back to Exercise 7.1, when (i.e. at what distance) is the wind speed back up at 10 m/s?*

7.11 *Looking back at Figure 7.12, determine the cut-in and cut-out wind speeds and the rated power and speed of the generic wind turbine.*

7.12 *Calculate the cost of a 5% wake loss for a big offshore wind farm. This is a quite open question of course, but try to make some reasonable assumptions.*

7.13 *Do a search for 'deep array effect' and see what the current thinking within the field is.*

8

Modelling

Once you have finished reading this book, it is very likely that you will only see the topics described here through the 'lens' of various models, that is, you will get some data, put it into a model and then get some output, or you might just get the output from someone else. Therefore we will address this situation explicitly in this chapter. This will be a more philosophical chapter, addressing modelling, the nature of it, input, output, errors, when to use and when not to use a model, what a good model is, etc. I am unfortunately not a philosopher, but I will do my best to derive some general conclusions and state some general remarks.

8.1 Modelling and What it Means

To start out we will define what we mean by a model. Models can mean a whole lot of different things, but when we talk about the term here, we mean specifically a *mathematical* model, which can be defined as:

Mathematical model: a description of a system using mathematical concepts and language.

So, most of what we have discussed in most of the chapters have been mathematical models. Schematically a mathematical model can be described as:

$$\boxed{\text{input}} \Longrightarrow \boxed{\text{model}} \Longrightarrow \boxed{\text{output}}$$

This is also the structure we will use in our discussion of modelling. When looking at the model, the most important question to ask is:

How good is the model for the task at hand?

'Good' can mean and imply a lot of different things. Here we will focus on accuracy and validation, but also appropriateness will be looked at.

Meteorology for Wind Energy: An Introduction, First Edition. Lars Landberg.
© 2016 John Wiley & Sons, Ltd. Published 2016 by John Wiley & Sons, Ltd.

8.2 Input

We will start with the input to the model. Looking back at all the chapters in the book we have seen various forms and types of input: time series data: measurements of wind speed, temperature, pressure, stability, etc.; information about roughness either as a single value for a sector (in the case of simple profiles) or a map of the roughness of the site. Input could also have been the description of an obstacle or an orographic map of the site.

What all these have in common is that they are numerical descriptions of some thing 'out there' in the real world. There is an important point, which we will see very clearly when we talk about chaotic systems, but even for non-chaotic systems. It is important to know how *accurate* the numerical data we put into the model is. Please refer back to Section 3.1 for a much more detailed discussion of this. Because, if the data we input is not accurate, even the best model cannot produce the right answer. This fact is also known as the GIGO principle: garbage in – garbage out. In Box 23, I have given a very simple example of what happens when there are errors in the input.

The other aspect of the input I would like to highlight here is how *detailed* the input is. As an example consider a very sophisticated flow model which you want to use to model the flow over a big area. To get the most out of the model we would need input that matched the sophistication of the model; this could be a high-resolution description of the landscape, a number of measurements with a high vertical resolution, etc.

8.3 Modelling

The central part of the input–model–output chain is of course the modelling part. Discussing modelling there are a few important questions that need to be asked[1]:

- Do I understand the problem?
- Do I understand the model?
- Is the model at the right level of sophistication?
- Do I have the input to match the sophistication of the model?
- Do I have the physical resources to run the model?
- Do I understand the output of the model?
- Do I understand the accuracy of the output?

These questions are quite related of course, but we will discuss each of them in turn in the following.

Box 23 Effect of errors in the input

As a very simple example, imagine that we have a model saying that the output is the square of the input, viz

$$m(x) = x^2 \tag{8.1}$$

---

[1] The 'I' in the questions should of course be taken to mean your organisation, not necessarily you personally.

where m is the model and x the input. We are now looking at the case where the real value of x is 10 (and $m(10) = 100$). To see the effect of accuracy we will calculate four examples with the following accuracies:

- ± 10, that is, the same order of magnitude as the input
- ± 1, 10% of the input
- ± 0.1, 1% of the input,
- ± 0.01, 0.1%.

I know you can fairly quickly see what the consequences are, but here are the results:

- ± 10: $x \in [0, 20] \rightarrow m \in [0, 400]$: 100–300% error
- ± 1: $x \in [9, 11] \rightarrow m \in [81, 121]$: a 20% error
- ± 0.1: $x \in [9.9, 10.1] \rightarrow m \in [98.01, 102.01]$: a 2% error
- ± 0.01: $x \in [9.99, 10.01] \rightarrow m \in [99.8, 100.2]$: a 0.2% error

As you can see, the possible range of output is very dependent on the accuracy of your input, up to 300%, and about double of the accuracy.

I hope this very simple example has given you a feel of how the sensitivity of the output of a model can be on the input. Had I chosen a linear function, the effect would of course have been much less dramatic, but similarly, had I chosen a cubic function, which for example, the kinetic energy follows, the result would have been even more dramatic.

The first question 'Do I understand the problem?' is quite philosophical in nature of course, but it also has some very practical consequences, because, if the problem is not understood deeply and completely enough, it is difficult to choose the right model. As an example, imagine you want to know the wind at hub height at a particular wind farm site. Is it enough just to estimate based on what the wind is known to be in the area, or is it necessary to take the measurements from a nearby mast (maybe at a different height), do I need to use the wind profile to calculate the wind at hub height, or do I need to employ a full-fledged flow model to take the complexity of the surrounding terrain into account? The answer to the question will depend on what you want the result for: is it for example, just a screening exercise where you are choosing between dozens of sites, or are you at the final stages of deciding on the financing of a wind farm.

The next question 'Do I understand the model?' is possibly the most generic, and in many ways it is an umbrella question that spans the others further down the list, but it also goes deeper. Issues that need to be addressed here could include the proper academic background, training and also experience.

Once the problem is understood, a list of possible models can be identified, and the next question could be: 'Is the model at the right level of sophistication?'. Understanding the problem will help answer this question, but, the following questions do also play an important role. There are two sides to this, actually: the one side is to match the problem, but the other is circling back to the previous question, because if you do not understand the chosen model, it is very hard to use it in the appropriate way, and a simpler model, which you understand, might be a much better solution. Note also, that in many fields within wind energy the list of available models changes quite often, so make sure you have identified all the currently available ones.

The model has now been chosen and the following questions need to be answered. The first one 'Do I have the input to match the sophistication of the model?' goes hand-in-hand with the previous question, very limited or inaccurate input will probably not be a good match for a very sophisticated model, a more simple model might be equally good or even better, since the sophisticated model would try to specialise in the inaccurate data, producing inaccurate and falsely detailed output.

Everything up until now might have pointed to a specific, maybe very advanced, model; however, if the question 'Do I have the physical resources to run the model?' is not answered satisfactorily, you are getting nowhere fast. The traditional hierarchy has always been that the PC is tried first, i.e. is there enough processing power, RAM, storage etc available on my PC to run the model, and can it be done in a reasonable time? Clearly as the computational requirements of the models increase (even compared to the fact that the processing power of the PCs also increases) this is getting more and more difficult, and main frames, super computers, etc. will need to be used. Recently (looking from 2015 at least) a dramatic change has happened and that is the so-called *cloud computing*, like Amazon's AWS (Amazon Web Services, Amazon, 2015) and Google's, Google Cloud Platform (Google, 2015). These are very very powerful computers accessible to anyone via the internet, this means that virtually everybody has a super computer with unlimited storage at their disposal. This of course also means that we need to extend the question about understanding the model to also include a question about one's ability to operate mainframes, super computers and cloud computing.

The problem is now understood, the model selected, and maybe run on a fancy cloud-based computer and now the next question needs to be asked: 'Do I understand the output of the model?'. If for example, the model is a simple wake model, or maybe the wind profile, the chances are that, yes, I do understand the model. The biggest danger I see under this heading is the output from very sophisticated models: it can be very detailed with vast amounts of information produced (millions of numbers), look very accurate (maybe due to the high resolution), and be displayed very nicely (three dimensional and in colours!), and still the people looking at the output might be misled, they might not know what they are looking at, is it an instantaneous value or one averaged over many months? What exactly is the quantity being looked at? And going back to the previous questions, it might be based on very inaccurate input data, rendering the output nicely looking but useless! An element of this question is also: am I – through this understanding of the model – able to interpret the output correctly?

Which nicely leads us to the last question: 'do I understand the accuracy of the output?'. This will be discussed further in Section 8.5, but basically, you need to understand both of the two types of errors you can encounter.

Having gone through this long list of questions enables you to get a deep understanding of the problem and the model; sometimes it will be very easy to answer the questions, others not easy at all. The easiness of course scales with your level of expertise!

To finish off the discussion about the modelling part of the chain, an example of a very complex and sophisticated model, with a lot of relevance to wind energy, will be described

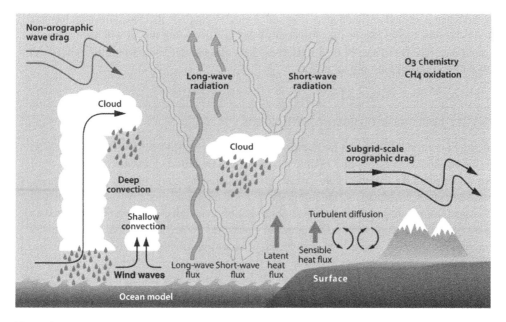

Figure 8.1 Some examples of the numerous sub-grid processes that a NWP model needs to take into account. Source: © ECMWF. Reproduced with permission of ECMWF under the Creative Commons Attribution-Non-Commercial-No-Derivatives-4.0-Unported Licence

in some detail in the following. We are going to address the so-called Numerical Weather Prediction (NWP) models.

8.3.1 Numerical Weather Prediction Models

In this section, the NWP models will be described. The models normally called NWP models are models covering flow at least on the synoptic scale, but also all the way up to the global scale. Examples of some of these can be found in Box 26. As with other flow models, the models use a three-dimensional grid for the calculations (c.f. Figure 8.1).

Exercise 8.1 *What were the characteristic time and length scales of the synoptic and the global scale?*

When running a meteorological model in an operational setting, the process follows the following four steps:

1. observation
2. assimilation producing the analysis
3. prognosis
4. post-processing

The first step is the observation step where all kinds of data is collected. This step has to be taken before the model can be run. A modern weather forecasting model will have a long list of input data (see Box 24). All these different kinds of data will come from different sources with different resolutions and accuracies.

Box 24 Input data and sources to NWP models

In order to run an NWP model we need to input the initial (starting) state of the atmosphere, so the model can progress this as time passes. The state of the atmosphere is made up of the values of all the variables that we use in the model, that is, pressure, p, temperature, T, wind velocities, u, density, ρ, and humidity, q (more advanced models would also include chemical compounds and nucleation agents as dust and salt). In principle, we need to know the value of each of these variables in every grid point of the model to initiate any simulation. This is of course not possible, but observations of all of the above variables are obtained from the following sources:

satellite-based instruments, ground- and ship-based weather stations, buoys, weather balloons, aeroplanes and other available sources.

In order to swiftly collect, exchange and communicate this huge amount of data, the world's meteorological institutions, through the WMO (World Meteorological Organisation), have developed the Global Telecommunication System (GTS), which does exactly that (WMO, 2015).

This means that the raw observed data is actually quite a mess, and were we to take all this data and input it directly into a numerical model, the model would be quite confused, since the picture painted by the data can be very internally inconsistent and, importantly, not in agreement with the physical laws represented by the equations in the model. This means that, if the model is just run with the raw input data, the outcome would be very unphysical many time steps into the future, so many in fact, it would be useless or of very limited use. This is why the second step in the model chain has been inserted between the raw observations and the running of the model.

During the assimilation/analysis step, the raw data is adjusted in such a way that the entire fields of for example, pressure, wind speed, etc. are in agreement with the physical equations the model assumes they are. As a simple example, imagine a pressure field where the observation from one location would result in a very strong local gradient, say, because of inaccurate observations and not because of the observation of a real physical phenomenon. The model would then have to go through the motions and calculate how this gradient would develop and eventually fizzle out. Resulting in model calculations that are physically incorrect and useless for weather prediction. So, by taking the assimilation/analysis step we end up with a physical field that looks right to the mathematical model; this field is called the *analysis*. There is another benefit of this step, and that is the so-called reanalysis data sets (see Box 25 for more details).

Box 25 Reanalysis data

As we saw above, the assimilation/analysis step is part of getting the data to fit the model in such a way that the model does not get confused by the input it gets from the real world. The schemes that are used to do this, are very advanced mathematical models in their own right. As you can imagine, these schemes have developed with the models themselves over the years, and a new scheme will be better than an old one.

To study how the climate changes, it would be very valuable to have a consistent set of these analyses ranging back as long as possible. This is exactly the point of the so-called re-analysis projects: these projects take all the observations ranging tens of years back and put them though one analysis scheme. This will then produce a consistent set of analyses (i.e. initial conditions for NWP models), which can be used to see how the climate has evolved over long periods of time.

In wind energy, reanalysis is used for many different applications, but by far the most used one is as input to the so-called Measure–Correlate–Predict (MCP) method:

MCP is a method that correlates the (short) time series measured at the site (step 1), with the (long) time series extracted from a reanalysis data set (step 2). Once the correlation between the on-site data and the long-term data has been established, it is possible to create a synthetic time series of what the wind would have been at the site, had there been measurements for a much longer period (step 3). This enables us to determine the long-term statistics (especially the mean wind) of the site.

The set of reanalysis data sets available is growing by the day (mainly spurred by climate change applications) and it includes:

- ERA interim: 1979–present (Uppala et al., 2005)
- MERRA: 1979–present (Rienecker et al., 2011)
- NCEP-NCAR (older): 1948–present (Kalnay et al., 1996)

But there are many many more (see Exercise 8.5).

Once the observations have been input to the model (assimilated) and 'beaten into submission' by the analysis, we are ready to start the actual calculation, and since this is calculating a state into the future, this is called the *prognosis*.

In a very simple way, what the model does is that it takes a very-high-resolution grid of the current status/value of all the variables that we looked at in the box covering the entire model domain, and then calculate what the value would be one time step ahead based on the physical laws represented in the model.

Which mathematically looks as follows:

$$V(x, y, z, t + \delta t) = f(V(x, y, z, t)) \tag{8.2}$$

where $V(x, y, t)$ is the value of one of the variables, at location (x, y, z) in space and time t, and f is the atmospheric model, and δt represents the time step.

This is of course a very complicated thing to do, which is why all meteorological institutes/offices have lots of clever people and very powerful supercomputers (often the met institutes' computers can be found among the world's fastest computers, see e.g. top500.org). The more powerful the computer, the more grid points, and the more sophisticated (and thereby computationally more demanding) models can be used to model the weather, resulting, hopefully, in more accurate results and forecasts that can predict further into the future. The result of the prognosis is therefore a whole lot of values in a very detailed grid spanning the globe, region, country, area that the model is set up to forecast for, for each time step in the calculation.

To illustrate, analyse and communicate all these values, the final step, the *post-processing*, is required. In the post-processing step the weather maps we know from the telly are some of the best-known outputs, but if you go to any meteorological institute you will find numerous and sometimes very sophisticated other types of output, like probability weighted precipitation and temperature forecasts based on ensemble runs (see Section 8.7.1).

Exercise 8.2   *Go to the website of your local meteorological Institute/office/centre and find examples of output from a meteorological model that has been postprocessed to illustrate the output of a numerical model.*

You can now see that understanding and evaluating the quality of a numerical weather prediction model requires that you understand the data that has been assimilated into the model, how this data has been preprocessed in the analysis part, and once all this has been done, you can evaluate the model itself. There is a very long list of different kinds of NWP models, and I have listed some of the important ones in Box 26.

Box 26   Some NWP models

Different NWP models around the globe:

- GFS (Global Forecast System), NOAA/NCEP, (NOAA, 2015a)
- ECMWF (European Centre for Medium-Range Weather Forecasts), (ECMWF, 2015)
- Met Office Unified Model, UK Met Office, (UK Met Office, 2015)
- HIRLAM (High-resolution limited area model), Danish Meteorological Institute (DMI), (DMI, 2015)
- Global Spectral Model (GSM), Japan Meteorological Agency (JMA), (JMA, 2015)

You will find that, as we discussed in the chapter on local flow, justification for sophistication of a model can be based on numerous arguments, including economic, skills and, I am sure, also political ones.

8.3.2 Sub-Grid Processes

All numerical models are based on having the values of the quantities they model (wind, temperature, pressure, etc.) represented on a grid. A grid is defined by the number of points in the three dimensions (x, y, and z), and the distance between the grid points is a measure of the resolution. The size of the grid depends on the area the model is applied to, a big area will have a longer distance between the grid points and a smaller area a shorter distance. A high-resolution model will then have relatively more grid points, and thereby a shorter distance between the points, and vice versa for a low-resolution model.

The higher the resolution, the more memory and computational power is required, so there is always a trade off between the available computational resources and the resolution.

No matter the resolution, there will always be physical processes that happen on scales that cannot be resolved by the chosen grid, these are called sub-grid processes (see Figure 8.1). In some cases these can just be ignored, but in others, it is important to include their effect, and a very essential aspect of model development is to identify the important sub-grid processes, and develop models (called *parameterisations*) that take their effect on the flow being modelled into account.

In some cases, a very big area needs to be modelled but there can also be a smaller area (or several smaller areas) that are of particular interest. A trick is to refine the resolution of the model in just these areas, this technique is called *nesting* (see Figure 8.2). Nesting means that within the model domain, grids of higher resolution are inserted, producing output in these areas also of higher resolution, whereby the accuracy of the model is improved in the

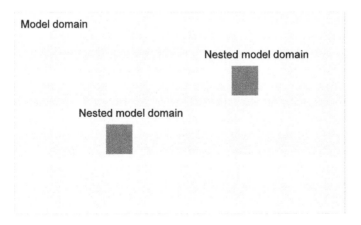

Figure 8.2 A simple illustration of nesting, where two areas of higher resolution have been inserted in the overall model domain

areas of interest; knowing whether this has taken place is important, especially when using and validating the model in a specific point.

8.4   Output

The last link in the input-model-output chain is the output. Model output can be many very different things, ranging from a single number, as would be the case if the model was the log profile, and we were modelling the wind speed at a certain height, to many gigabytes of data if we are looking at the output from an NWP model (as mentioned above). No matter what it is, we need to know exactly what the output represents – in the case of the log profile, we would need to know, for example, what is the averaging time of the wind speed, is it an instantaneous value, a 10-minute averaged value or something different altogether. In the more complicated case of the NWP, the number of questions that need to be asked is of course many times larger, and they include the aforementioned averaging time, but another crucial question to ask is where on the surface does the value refer to? As we have discussed, an NWP model calculates all its values on a grid, however, some values are referenced to the grid point, others to the surface between the grid points and others again to the centre of four points.

8.5   Errors

There are two types of errors: inherent model errors and input dependent errors. Model dependent errors can be thought of as resulting from the fact that the model does not describe the physical system accurately. In wind energy as well as in many, if not most, other fields, it is fair to say that all models have this type of error. Likewise, as we discussed in the section on input, there will always also be some sort of error or uncertainty in the input, which will be propagated through the model all the way to the output. We have already seen this in Box 23 in Section 8.3, where even if the perfect model was x^2, we would still get wrong answers if the input was not accurate.

It is therefore very clear, that we will have to deal with errors when we deal with modelling. First, understanding how errors are propagated through the system is an important step in any type of modelling. We will shortly discuss chaos, but assuming that the system is non-chaotic, it is possible to do a so-called sensitivity analysis. In a sensitivity analysis you see (surprise, surprise!) how sensitive the model is to changes in the input. There are typically two tests you might want to do: one is to vary the input with what you expect the accuracy to be, and the other can be used if you are unsure of a certain number, for example, the roughness, then you would run the model with the different possible values, and see how sensitive the system is to this variation. Sensitivity analysis works great for simple systems where the inputs are limited, but as the complexity of the model increases, one needs to develop strategies for which inputs to vary to get the best measure of the sensitivity.

The model errors is the other type of errors that can occur. Imagine (see Figure 8.3) a very simple case where the real phenomena is described by the square of the input, and the model you have at hand is linear in nature. It is impossible to fit a linear graph to a square function, so in order to be able to deal with such a model, you need to know (in the example here) that

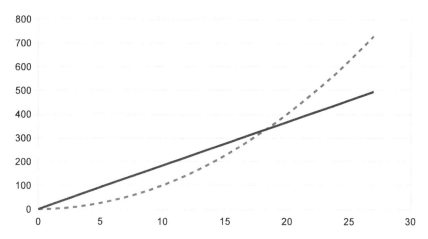

Figure 8.3   A linear function (solid line) used to model a square function (dashed line)

for lower values the model will over-predict and for higher values it will under-predict. You can do a little bit more, too, by fitting the model to the range of values you are expecting the model to operate within, this will minimise the errors, but of course not eliminate them.

This is indeed a very simple example, but the principles of knowing what kind of behaviour to expect from the model and of fitting it to the ranges of input you expect can easily be generalised to much more complicated models. It will be a more difficult undertaking of course.

It might not be a general fact about modelling, but at least in wind energy, and in particular with the modelling of flows, we have seen in practice, that time and time again, we are saved by the fact that the errors tend to cancel each other out. As an example of this 'luck' consider a neutral model, used in a case where stability plays a significant role. If we, to a first approximation, can assume that the effect of stability is symmetric, and there are as many stable as unstable events, then there is a tendency for these errors to cancel each other out (under-prediction cancelling out over-prediction), actually resulting in quite accurate model predictions. As my former colleague Søren Ott at Risø used to say: the model is right for all the wrong reasons! We are of course not always saved by this fact, and again knowing when we are and when we are not, will make the whole difference.

8.6   So, What is a Good Model?

If you have read this far, the answer or at least the contours of the answer should be clear. Going back to the list in Section 8.3, a good model is one where you can answer yes to all the questions on the list:

- Do I understand the problem?
- Do I understand the model?
- Is the model at the right level of sophistication?

- Do I have the input to match the sophistication of the model?
- Do I have the physical resources to run the model?
- Do I understand the output of the model?
- Do I understand the accuracy of the output?

If you have thought of a specific modelling situation when you went through the questions, it is far from certain that you were able to answer 'yes' to all the questions. Many a time the answer is not 'yes' nor 'no', but 'it depends'; however, the more you know about the model and the situation you want to model, the better you can qualify what it depends on.

There is a further aspect of what a good model is and that has to do with *validation*. Validation means that you compare the output of your model to some high-quality measurements. By doing this, you not only get an idea of how well the model performs, but you also get an idea of the errors and accuracy in connection with the modelling of this particular case. For validation to be as good as possible, independent validation is of course preferred; independent validation means that it is not the developer of the model that carries out the validation and this of course ensures a certain amount of objectivity in the analysis of the results. Another much used way is to do so-called blind tests. In these the input that describes the situation are given, but not the corresponding output. An example could be a complete description of a site, the task is to predict the wind speed and direction at a certain number of locations, where measurements have been taken, but the measurements are not shared until after the predictions have been handed in.

In wind energy, we currently see an increase in the number of published validation studies both independently but also by people who have developed and run the model. With an increasing amount of evidence for how well the models perform, we are able – as an industry – to get a better view on how well the models perform in different situations.

An essential part of validating a model is the comparison to measurements, and that, unfortunately, is not always that easy. A very thoughtful analysis can be found in Ayotte et al.'s paper (2001), and my own take on this is that it is not easy, since we almost always measure at one (or a few) point in space and time, and comparing this quite limited information to the plethora of data you get as output from a model requires a good systematic approach and a careful analysis of what is actually compared.

8.7 Chaos

It is not possible to discuss modelling, and especially not atmospheric modelling, without a discussion of chaos. The kind of chaos we will discuss here is of a very special kind called *deterministic chaos*, which basically means that we can calculate how chaotic a system is.

The basic problem is that chaotic systems are hypersensitive (to use a not-so-scientific term) to the input given to them, and since we can never measure with 100% accuracy, we have the problem. Since we in general, on our human scale, live in a non-chaotic world, where if we, say, throw a ball in the same direction several times with more or less the same force, expect it to land in more or less the same area – we therefore get very confused when we meet chaotic systems where this is not the case. Had throwing-a-ball been a chaotic system, the balls could have ended up in places very far away from each other in the example above even though the throws were almost identical.

This has some profound effects on weather forecasting, since the atmosphere *is* a chaotic system. Richardson (see Section 6.2) and many of his contemporaries knew that the atmosphere was a complicated system, but he thought that if we had enough (at the time human) calculators (see Wired, 2015, for a recent description of this) we would be able to calculate the weather well into the future. It was not until Lorenz in the early 1960s (Lorenz, 1963) happened to stumble upon (so the story goes anyway) a case where two calculations with seemingly the same input, gave two very different outcomes. When he looked closer at the input, he realised that the two sets were in fact different, but only ever so little – chaos had been discovered.

The chaotic nature of atmospheric flow has been popularly described as when a butterfly flaps its wings in Brazil, it could lead to a tornado in Texas (and that was exactly as Lorenz put it in his paper, Lorenz, 1972). I think this has led to the slight misconception that chaotic systems are also explosive, wild or dramatic, this is not the case in general. Chaos just means that a small change (the butterfly flapping its wings) changes the state of the atmosphere and the path it will follow will very quickly be very different from the path it would have followed, had the butterfly not flapped its wings at that particular time and place. This story is also the reason why chaotic behaviour is also said to be called the butterfly effect. In Box 27, I have described a very simple but also very famous chaotic system resembling the atmosphere, the Lorenz attractor.

Box 27   Lorenz attractor

The Lorenz attractor is the plot of the trajectory of the solution to the following set of differential equations, which can be seen as a very simple model of a two-dimensional fluid flow.

$$\dot{X} = \sigma(Y - X)$$

$$\dot{Y} = -XZ + rX - Y$$

$$\dot{Z} = XY - bZ$$

where the dot over the variables refers to the time derivative, varying r changes whether the system is chaotic or not.

The solution to one of the chaotic cases can be seen below.

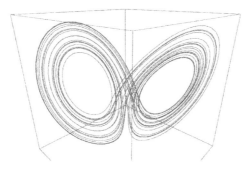

The chaotic nature means that if two particles started very close to each other, they would very quickly find each other very far away from each other, for example, on opposite 'loops'.

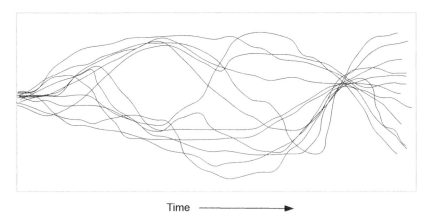

Time ───────────▶

Figure 8.4   Illustration of an ensemble prediction. There are points in time where the system seems to fall in one of two stages, others where it is very unclear, and at one time all the paths go through the same state. Time is along the *x*-axis, and a generalised state (think e.g. of wind speed) of the system along the *y*-axis

Knowing that the atmosphere is chaotic very quickly begs the question: how can we predict the future weather with any kind of accuracy? How can we say anything at all about such a chaotic system? The answer to these questions is that we can't directly; however, we can get some very useful information in an indirect way; through so-called ensemble prediction.

8.7.1   Ensemble Prediction

Ensemble prediction works in the following way: the same model is run many times (say 100) with the input changed just a little bit for each run. This produces a big number of possible outcomes of the system, and the clever thing is that looking at all these statistically, we can gain insight into the *probabilities* of the different types of outcomes. I have tried to illustrate this in Figure 8.4, where you can see that even though the outcome can be quite different, we can see that there are times when we know with some certainty what a limited set of possible outcomes might be. At the same time, we also get a feel for how certain the prediction might be. Because if the different paths vary widely, then it is very likely that the prediction of this situation will be uncertain, whereas if they all follow more or less the same track the prediction will most likely be more accurate.

8.8   Summary

In this chapter we have discussed modelling from a more general perspective. We used the input → model → output chain to discuss and analyse the individual parts.

Looking at input, we identified three important aspects: how representative, accurate and detailed is the input data in question. You will recognise many of these aspects from the discussion of measurements, which of course is not so strange since a lot of input is actually measurements.

The central part, the modelling, tried to cover the subject from a general point of view by asking: what is a good model? A long list of sub-questions very naturally appeared, and by answering those, we put ourselves in a good position to answer the main question.

We discussed the importance of understanding the error, and looked at a very simple, but illustrative, example of a model. It is fair to say, that errors, and especially unknown errors, can lead to very wrong results and conclusions.

I think that I somewhere promised you that there would be no new models in this chapter, that was almost true, but I did have to describe the Numerical Weather Prediction model is some detail. These models are very important for modelling the weather as described in Chapter 2, which is the reason why I have included them.

The final link of the chain, the output, was then discussed, and it was shown how important it is to understand the input and the model to really understand and use the output.

The last part of the chapter was about chaos! The main equations that govern the atmospheric flow are chaotic, so, trying to understand this aspect, helps us to understand the modelling. A 'solution' to the fact that the atmospheric system is chaotic, is the so-called ensemble methods, and we also discussed them briefly.

Exercises

8.3 *In Figure 8.4 identify the points in time where the system seems to fall into one of two stages, where it is very unclear, and the one time all the paths go through the same state.*

8.4 *Find the name and description of a few more NWP models.*

8.5 *Find a few more re-analysis data sets.*

9

Conclusion

You have now come to the end of a very long journey, and to wrap up I would first like to briefly go through the various topics we have covered and then close with a few concluding remarks and of course finish with some exercises, the last one intended to send you off to learn more on your own.

We started out by looking at what can be termed general meteorology. Here we covered the variables, the forces, some common types of weather and much more, the purpose was to give you an introduction to those fundamentals of the weather that have an influence on wind energy, and also to give you a broad foundation in meteorology.

We then looked at measurements from a theoretical and sometimes philosophical angle, where we covered a discussion of why we measure and what we want to learn by measuring, we looked at some of the analysis tools needed to understand time series. The last part of the chapter described a long list of instruments that we use in meteorology, again with a special emphasis on the ones we use for wind energy purposes.

After the chapter on measurements, we had an understanding of what governs the 'weather' and how to measure it, and we then continued by zooming in on some of the more specific topics. The first was the vertical profile, mainly of wind speed, but direction was also covered briefly. We started out by 'deriving' the first of the fundamental laws in wind energy meteorology: the logarithmic wind profile in its neutral version. We then continued to complicate things by introducing zero-plane displacement, and also stability, after a relatively detailed description of what atmospheric stability is. An important connection was made between the wind at the surface and the wind aloft: the geostrophic drag law, which was already introduced in the chapter on weather.

In order to understand the flow at a site further, we then turned to describing the local flow, ideally speaking, one 'perfect' model should cover all the various local flows, but since such a model does not really exist, we turned to separating the flow at the site into a number of local flow types with different causes: orography, roughness, obstacles and thermally driven. Simple models and rules of thumb can only take us so far, so we also discussed more advanced models like the so-called CFD models, which attempt to model flow when it is separated, a flow type we often see in complex terrain.

Meteorology for Wind Energy: An Introduction, First Edition. Lars Landberg.
© 2016 John Wiley & Sons, Ltd. Published 2016 by John Wiley & Sons, Ltd.

We then turned to two more advanced topics: turbulence and wakes. In the chapter on turbulence, we first had to start thinking in a slightly different way about how we understand a phenomenon, since turbulence is one of the 'unsolved' problems in physics: we cannot model it directly, but can talk about it in various round-about and more statistical ways. One of the most important tools is the spectrum, which was introduced. We also looked at loads which is a cousin of turbulence.

The last specific topic we covered was wakes. We looked at three different relatively simple but powerful models for wakes, we looked at how the wakes of several wind turbines in a wind farm interact and how to model that, and finally looked at some more advanced issues like wakes in very big wind farms, stability dependence and how two wind farms affect each other.

The final chapter was dedicated to modelling, modelling more as a general topic, than addressing a specific model – we introduced a list of questions that need to be asked in order to fully benefit from and understand a model. I did manage to sneak one more model in: the Numerical Weather Prediction model, which is the type of model that predicts the weather. In some way the circle is therefore closed, since we started and ended with the 'weather'.

As a general principle, all through the book, I have tried to cover the basics in an as simple way as possible, but always using maths to explain, and then explaining the maths. I have tried to simplify things as much as possible, which has meant that I have cut some corners, of course, but more importantly, I might have given you the impression that meteorology for wind energy is simple! It is not, and in real life there will always be complications making the situation not so straightforward. However, with the knowledge you have obtained from this book you should be able to say: 'it depends' and then also be able to list and discuss the multitude of aspects 'it' depends on.

To close the book I would like to set you two final exercises (I very much hope you, by now, also have solved most of the exercises in the various chapters), one looking at the 'it depends' question and a final one to make sure you continue to learn.

Exercise 9.1   *Returning to Section 3.1 put together a list of things that need to be taken into account when siting a meteorological mast on a site. (I have put elements of an answer in the Answer's appendix)*

Exercise 9.2   *This is the final exercise: take this book, go to a site, set aside 20 minutes to look at and identify as many of the topics we have discussed in the book. Repeat a few times over the next year, and you will not only have read the book, but also learnt and remembered something!*

A

Cheat Sheet

Kinetic energy

$$E_{kin} = \frac{1}{2}mu^2 = \frac{1}{2}(\rho u)u^2 = \frac{1}{2}\rho u^3$$

The variables

- Pressure, p, (Pa)
- Temperature, T (K or °C)
- Density, ρ, (kg/m^3)
- Velocity, u, v, w (m/s)
- Humidity, q, (g (water vapour)/kg (dry air))

The forces

- The pressure gradient force
- Frictional force
- Coriolis force
- Gravitational force

The layers of the atmosphere

Name	Start height (km)	Pressure (hPa)	Temperature (°C)	Comments
Troposphere	0	1013	Falling, 20 to −50	This is where we live
Tropopause				Jet stream
Stratosphere	11	226	Raising, −50–0	Temperature increases with height due to the presence of ozone
Stratopause				
Mesosphere	47	1	Falling, 0 to −90	

(continued)

Meteorology for Wind Energy: An Introduction, First Edition. Lars Landberg.

Name	Start height (km)	Pressure (hPa)	Temperature (°C)	Comments
Mesopause				
Thermosphere	85	≈0	Raising, −90 −	The space shuttle flew here
Thermopause				
Exosphere	700	≈0		
Exopause				

Time and length scales

Name	Length scale (m)	Time scale	Phenomenon
Global/planetary	10^7	O(weeks), 10^6 s	Planetary waves
Synoptic	10^6	O(days), 10^5 s	Cyclones
Meso	10^5	O(hours), 10^3 s	Sea breezes
Micro	10^2	O(minutes to hours), 10^2 s	Thunder showers

Instruments

Instrument	Measures	Instrument	Measures
Cup anemometer	Wind speed	Psycrometer	Rel humidity
Wind vane	Direction	Barometer	Pressure
Sonic	Speed	Lidar	Speed
	Direction	Sodar	Speed
	Temperature	Ceilometer	Height
Hot wire	Speed	Scatterometer	Wind speed
Pitot tube	Speed		Direction
Thermometer	Temperature		

The logarithmic wind profile (neutral)

$$u(z) = \frac{u_*}{\kappa} \ln\left(\frac{z}{z_0}\right)$$

Geostrophic drag law (neutral)

$$G = \frac{u_*}{\kappa} \sqrt{\left[\ln\left(\frac{u_*}{fz_0}\right) - A\right]^2 + B^2}$$

The four types of local flow

- Orography
- Roughness
- Obstacles and
- Thermally driven

IBL 1:100 rule

An internal boundary layer grows as 1:100.

Stability

Name	Quantity	Stable	Neutral	Unstable
Lapse rate	Γ	$< \Gamma_d$	$= \Gamma_d$	$> \Gamma_d$
Potential temperature	$d\theta/dz$	> 0	$= 0$	< 0
Heat flux	H	< 0	$= 0$	> 0
Monin–Obukhov	L	> 0	$= \infty$	< 0
	z/L	> 0	$= 0$	< 0
Smokestack (Box 12)		Fanning	Coning	Looping

Wakes

NO Jensen model for reduction of wind speed in wake

$$u_w = u_i \left[1 - (1 - \sqrt{1 - C_t}) \left(\frac{D}{D + 2kx} \right)^2 \right]$$

Turbulence intensity

$$\text{TI} = \frac{\sigma_u}{\bar{u}}$$

Roughness

Landscape type	Water	Snow	Grass	Farmland	Forest	City
Roughness (m)	10^{-4}	10^{-3}	0.03	0.1	0.8	1.0

Charnock's relation

Offshore roughness

$$z_0^w = a_c u_*^2 / g$$

Questions to ask of a (flow) model

- Do I understand the problem?
- Do I understand the model?
- Is the model at the right level of sophistication?
- Do I have the input to match the sophistication of the model?
- Do I have the physical resources to run the model?
- Do I understand the output of the model?
- Do I understand the accuracy of the output?

B

Answers to Exercises

Chapter 1 Introduction

1.1 Please see text following Exercise 1.1.

1.2 Please see text following Exercise 1.2.

1.3 Please see text following Exercise 1.3.

1.4 Please see text following Exercise 1.4.

1.5 Please see text following Exercise 1.5.

Chapter 2 Meteorological Basics

2.1 Please see text following Exercise 2.1.

2.2 Newton's second law of motion states that the acceleration a particle sees is equal to the forces acting on the particle normalised by the mass of the particle, or:

$$\vec{F} = m\vec{a}$$

2.3 It takes 24 hours ($=24 \times 60 \times 60$ s $= 86{,}400$ s) to turn once (i.e. 2π in radians), so $\Omega = 2\pi/86{,}400 = 7.272 \cdot 10^{-5}$ rad/s. As you can see, this is not quite the number given in the text, and the reason is that the Earth moves in its own orbit, too, so seen from far away (i.e. from a fixed frame of reference) the Earth turns faster. This 'day' seen from this reference is also called a *sidereal* day and is 23 hours 56 minutes and 4 seconds (this I must admit I did not understand when I read it first, but what convinced me was when I consulted http://en.wikipedia.org/wiki/Earth's_rotation for a bit more

Meteorology for Wind Energy: An Introduction, First Edition. Lars Landberg.
© 2016 John Wiley & Sons, Ltd. Published 2016 by John Wiley & Sons, Ltd.

of an explanation), so we need to divide by that number instead to get to the correct rotation.

2.4   The magnitude of the Coriolis parameter is:

$2\Omega \sin 0 = 0$ at the Equator

$2\Omega \sin 90 = 2\Omega \cdot 1 = 1.458 \cdot 10^{-4}$ rad/s at the North and South Pole

$2\Omega \sin 56 = 2\Omega \cdot 0.829 = 1.209 \cdot 10^{-4}$ rad/s where I live (in Denmark, at 56°N)

2.5   Taking the y-axis to be parallel to the isobars, we get:

$$P_x = \frac{1}{\rho}\frac{\delta p}{\delta x} = \frac{1}{1.225}\frac{10}{500} = 0.016 N/kg$$

and $P_y = 0$.

2.6   We have $u_* = 0.3$ m/s, latitude 40°N, $z_0 = 0.1$ m. First we will calculate the speed of the geostrophic wind:

$$G = \frac{u_*}{\kappa}\sqrt{\left[\ln\left(\frac{u_*}{fz_0}\right) - A\right]^2 + B^2}$$

$$= \frac{0.3}{0.4}\sqrt{\left[\ln\left(\frac{0.3}{2 \cdot 7.292 \cdot 10^{-5}\sin(40°) \cdot 0.1}\right) - 1.8\right]^2 + 4.5^2}$$

$$= 7.3 \text{ m/s}$$

and then the direction:

$$\sin\alpha = \frac{Bu_*}{\kappa G} = \frac{4.5 \cdot 0.3}{0.4 \cdot 7.3} = 0.462$$

leading to that the angle α is $\sin^{-1}(0.462) = 28°$.

2.7   There is no analytical solution to this question, so I have put the equation for the geostrophic drag law into a spreadsheet (with $G = 8$ m/s, latitude 55°S, $z_0 = 0.05$), and then iterated until I got $u_* = 0.316$ (to check I put that back into the geo drag law and got 8.00 m/s, which was the original number).

2.8   At first this might seem a bit odd, but remember that there is no friction at the top of the atmosphere, so there is nothing to 'hold' it as the Earth rotates through the empty universe.

2.9   The u and v components of the Ekman spiral for a geostrophic wind, u_G, of 9 m/s at height 300 and 1000 m at a latitude of 30°N can be found from:

$$u = u_G[1 - \exp(-\gamma z)\cos(\gamma z)]$$

$$v = u_G[\exp(-\gamma z)\sin(\gamma z)]$$

where z is the height above ground, $\gamma = \sqrt{\frac{f}{2K}}$, and K the eddy viscosity coefficient $\approx 1\ m^2/s$.

So, with $\gamma = \sqrt{\frac{2 \cdot 7.292 \cdot 10^{-5} \sin 30°}{2 \cdot 1}} = 0.00604$ we get

$$u(300) = 9 \cdot [1 - \exp(-0.00604 \cdot 300) \cos(0.00604 \cdot 300)]$$

$$= 9.35$$

$$v(300) = 9 \cdot [\exp(-0.00604 \cdot 300) \sin(0.00604 \cdot 300)]$$

$$= 1.43$$

for $z = 300$ m and

$$u(1000) = 9 \cdot [1 - \exp(-0.00604 \cdot 1000) \cos(0.00604 \cdot 1000)]$$

$$= 8.98$$

$$v(1000) = 9 \cdot [\exp(-0.00604 \cdot 1000) \sin(0.00604 \cdot 1000)]$$

$$= -0.01$$

for $z = 1000$ m.

Translating this into speed, s, and direction, d, we get:

$$s(300) = 9.46$$

$$d(300) = 8.7°$$

and

$$s(1000) = 8.98$$

$$d(1000) = -0.1°$$

2.10 In my case, I am located in Copenhagen. The microscale flow is therefore dominated by the small hills and valleys in the area. Copenhagen is located on an island (Zealand), and the mesoscale flow is dominated by the distribution of land and sea. Denmark is furthermore located in Northern Europe which sits in the middle of the synoptic-scale flow known as the westerlies, basically the track that most low-pressure systems follow, governed to some extend by the mountains in Norway. Finally, the location of Northern Europe, seen relative to the large land masses of Europe and Asia, and the large water body of the Atlantic sets the scene for the global/planetary flow.

2.11 What you see on the satellite picture depends of course very much on the weather the day you are looking. Use Figure 2.10 to identify the cold and the warm fronts, and then measure the distance across. This should give you an estimate of the length scale. It is possibly more difficult to determine the time scale, unless you are very lucky on the day you are looking, but you should be able to find one system that starts to develop, then reaches full maturity, and then quite quickly dissolves. Typically in a matter of days.

2.12 As with Exercise 2.11, use Figure 2.10 as a guide to what the different phases of the cyclogenesis look like. If you look at global images you should be able to find at least one low-pressure system on any given day.

2.13 As with Exercise 2.11, use Figure 2.10 as a guide to what the fronts look like on a satellite image. Most images have a time stamp, and by measuring the distance you should be able to infer the speed.

2.14 At the surface the air flows into a low pressure, which means that the air will raise over it, for continuity reasons.

2.15 Normally, high-pressure systems are very large (compared e.g. to a low-pressure system), and they move very little, as opposed to the low-pressure system, which will move quickly through the satellite picture. A high-pressure system can stay in the same location several days and sometimes weeks.

2.16 Use for example, Google to search for the latest on El Niño, La Niña and ENSO. Make sure to apply a grain of salt if the effects of global warming are mentioned.

2.17 Searching the internet you should be able to find the answer to this. Currently (March 2015) the El Niño has just arrived (according to NOAA), but it is quite weak, and expected to remain so.

2.18 Today, 22 March 2015, the NAO index is positive and around 0.5 (according to the National Weather Service, NOAA).

2.19 I have found the following: the Arctic Oscillation (AO), Pacific-North American Pattern (PNA), strongly linked to ENSO) and the Antarctic Oscillation (AAO).

Chapter 3   Measurements

3.1 Please see text following Exercise 3.1.

3.2 At first it might seem as if the resolution is the same as the precision, at 0.001 (three digits after the decimal point); however, one first notices that there are only even numbers as the last digit, which brings the resolution to 0.002. Looking even closer you can see that the numbers actually jump in steps of 0.532, which is the true resolution. The lesson here is very clear: one needs to look at data very carefully!

3.3 That would depend on how missing data was labelled. In some cases a blank line is inserted, which, depending on the plotting program, would either result in no line drawn, in which case you would see gaps in the time series, or a straight line across the missing interval, in which case you would see abnormally straight lines in the plot. If the values are labelled with a number, say –99.99, then you would see the line crossing over to the other side of the *x*-axis; a lot of missing data would lead to straight lines on the wrong side of the axis.

3.4 The density (i.e. the number of points per unit area) of the points represents the probability of the wind coming from a given direction with a given wind speed. More points mean that the probability is higher.

3.5 The data series we have used in this chapter is not climatologically representative because it is too short (only 3 months). All the seasons are not present, meaning that the annual variability is not in the data, and likewise, there is not a multitude of years, meaning that the inter-annual variability is not present.

3.6 Due to the shape of the cup, the wind blows around the cup if it comes from the 'back' of the cup, and 'into' the cup when it comes from the front, so in the picture it will rotate counter-clockwise. This makes the anemometer turn in only one direction. In very very weak winds, the cup can actually turn both ways, but as soon as the wind picks up, it can only turn one way.

3.7

Moving 5 m/s away from B				
T thrown	Position	Distance to B	Time to A	T when at B
0	0	10	10	10
1	5	105	10.5	11.5
2	10	110	11	13
3	15	115	11.5	14.5

so the frequency is 1/1.5 Hz

3.8 As a minimum you must: identify missing/bad data, calculate the mean and the standard deviation, and form the wind speed distribution. Plotting the wind rose will also be very useful.

3.9 This is of course quite a subjective question, but looking at a few different ones you will quickly, first, understand that they are different, and, second, also be able to distinguish wind (speed) from the other data types. If you cannot find some appropriate data, you can use the plot of the random data.

3.10 Looking at Box 6 and also on the definition of the terms in Box 5 you will need to address the characteristics of the instrument, if the underlying accuracy of the instrument is say ± 100, then it is clear that the numbers could very likely not be different, if the precision indicated by the numbers is actually also reflecting the accuracy, that is, ± 0.1, then the two numbers are clearly different. In between these two there is a sliding scale which at some point will change from the two numbers being most likely equal, to the two numbers most likely being different.

3.11 The mathematical expression for the Gaussian (also called the Normal) distribution is:

$$f(x, \mu, \sigma) = \frac{1}{\sigma\sqrt{2\pi}} \exp\left(-\frac{(x-\mu)^2}{2\sigma^2}\right)$$

where μ is the mean, and σ the standard deviation.

3.12 The two Weibull distributions (both with $A = 10$) are plotted below:

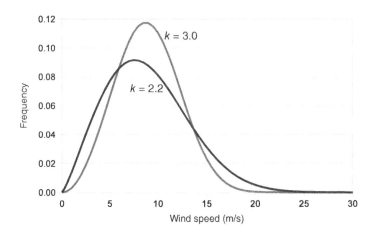

the k value is indicated next to the curves.

3.13 In order to get to Equation 3.5, we need to use Equation 3.4:

$$t_1 = \frac{L}{c+u}$$

and stating for the other way (note the minus):

$$t_2 = \frac{L}{c-u}$$

using the last equation to isolate c, we get:

$$c = \frac{L}{t_2} + u$$

and inserting this in the first equation:

$$t_1 = \frac{L}{\frac{L}{t_2} + u + u}$$

and isolating u, we get:

$$u = \frac{L}{2}\left(\frac{1}{t_1} - \frac{1}{t_2}\right)$$

which was the equation we were looking for.

3.14 This is because three lasers are used, each Lidar giving the radial velocity in the direction from the Lidar to the point. Thereby we have three components of a vector, but not in the correct coordinate system for atmospheric flow (i.e. the two horizontal and one vertical components). The three components can be transformed to that coordinate system using trigonometry. We therefore end up with the three-dimensional wind vector in the point the three Lidars are shining at.

Chapter 4   The Wind Profile

4.1 Please see text following Exercise 4.1.

4.2 Please see text following Exercise 4.2.

4.3 Please see text following Exercise 4.3.

4.4 Please see text following Exercise 4.4.

4.5 Please see text following Exercise 4.5.

4.6 Please see text following Exercise 4.6.

4.7 Please see text following Exercise 4.7.

4.8 Please see text following Exercise 4.8.

4.9 Looking at Equation 4.15 $u = 0$ when

$$0 = \frac{u_*}{\kappa}\ln\left(\frac{z-d}{z_0}\right)$$

which we can reduce to:

$$0 = \ln\left(\frac{z-d}{z_0}\right)$$

and again (using that $e^0 = 1$):

$$1 = \frac{z-d}{z_0}$$

which means that

$$z = z_0 + d$$

4.10   Please see text following Exercise 4.10.

4.11   Please see text following Exercise 4.11.

4.12

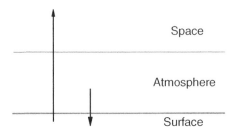

Simplified illustration of the nocturnal radiation balance between space, the atmosphere and the surface of the Earth. The arrows indicate the direction of the radiation.

4.13   The unit of L can be found from:

$$L = -\frac{u_*^3 \overline{\theta_v}}{\kappa g \left(\overline{w'\theta_v'}\right)_s}$$

by inserting the units of each of the quantities, viz

$$L = -\frac{(\text{m/s})^3 \ \text{K}}{[\cdot]\text{m/s}^2 \ (\text{m/s K})} = \frac{\text{m}^3 \ \text{s}^{-3} \ \text{K}}{\text{m s}^{-2} \ \text{m s}^{-1} \ \text{K}} = \text{m}$$

so a length.

4.14   Please see text following Exercise 4.14.

4.15   Please see text following Exercise 4.15.

4.16   This can be done in two ways, plotting the three points in a log–linear plot and then reading z_0 as the point where the line crosses the y-axis, and u_* the slope of the same line.

It can also be done mathematically, which is what I shall do here. You actually only need two points on a line to determine the slope and the offset, so we will use the third to check our calculations. It is quite useful to have the formulae for this, so I will start by deriving the general expression.

Assume you have two sets of values (z_1, u_1) and (z_2, u_2). Assume further that the wind profile follows the logarithmic profile. Find z_0 and u_*.

The log law says:

$$u(z) = \frac{u_*}{\kappa} \ln\left(\frac{z}{z_0}\right)$$

we want to isolate u_*:

$$u_* = \frac{u(z)\kappa}{\ln\left(\frac{z}{z_0}\right)}$$

inserting this in the original equation, we get (using the two points):

$$u_1 = \frac{u_2\kappa}{\kappa \ln\left(\frac{z_2}{z_0}\right)} \ln\left(\frac{z_1}{z_0}\right)$$

which after a good deal of algebra (please try!) can be reduced to:

$$\ln z_0 = \frac{u_1 \ln z_2 - u_2 \ln z_1}{u_1 - u_2}$$

we now have the two general expressions we need and are able to calculate first z_0 and then u_* (using the two highest points):

$$\ln z_0 = \frac{7.41 \ln(120) - 8.00 \ln(75)}{7.41 - 8.00} = -1.585$$

so $z_0 = 0.2$ m, calculating u_*:

$$u_* = \frac{8.00 \cdot 0.4}{\ln\left(\frac{120}{0.2}\right)} = 0.5 \text{ m/s}$$

Using the last point to check what we have derived, we get:

$$u(30) = \frac{0.5}{0.4} \ln\left(\frac{30}{0.2}\right) = 6.26$$

which was the measured value.

To illustrate this graphically I have plotted the profile and the three points below.

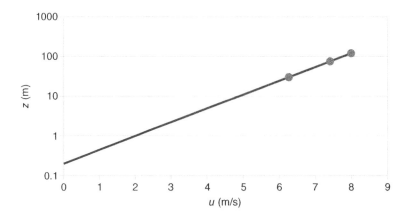

the points are indicated by the dots.

4.17 Using Equation 4.12 and inserting the values (75, 7.41) and (120, 8.00) we get:

$$\alpha = \frac{\ln(7.41/8.00)}{\ln(75/120)} = 0.162$$

4.18 Using Equation 4.11, the α from above and the 120 m wind (8.00) we get:

$$u(30) = 8.00 \left(\frac{30}{120} \right)^{0.162} = 6.38$$

which is quite close to the observed value (6.26).

4.19 Referring back to Section 2.3.3, you can see that once we get out of the surface layer, the direction of the wind starts to change and the speed starts to change behaviour, too, now following the Ekman spiral (see Section 2.3.3).

4.20 Using any of the values we can calculate the geostrophic wind, G (I have used a few more decimals of u_* in some of the calculations):

$$G(z_0 = 0.0002, u_* = 0.31) = 12.00$$
$$G(z_0 = 0.03, u_* = 0.43) = 12.00$$
$$G(z_0 = 1.0, u_* = 0.58) = 12.00$$

so, as you can see, I had used a geostrophic wind of 12 m/s.

4.21 It is true that the so-called greenhouse effect does indeed *raise* the temperature, relative to a planet without an atmosphere; however, at some point the radiation system reaches a new *equilibrium* and an equal amount of energy, to that received, is radiated back out to space, resulting in a higher, but stable atmospheric temperature.

4.22 The easiest way to solve this exercise is through trial and error. There are three variables so it is not straightforward but here is what I get after iterating (my initial guess was $z_0 = 0.4$, $u_* = 0.3$ and $d = 10$):

z_0:   0.8 m
u_*:   0.5 m/s
d: 20 m

You can also infer $z_0 + d$ from a log–linear plot, and if you then plot u against $z - d$, you can find u_*/κ as the slope of the line.

4.23 Calculating the profile for each of the layers (where the heights of the two changes are: 25.2 and 84 m), we get:

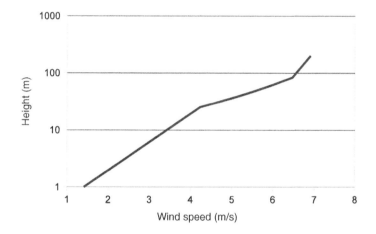

plotted in a log–linear plot.

4.24 Using Equation 4.17 and iterating, I get the height to be 131 m.

4.25 Calculating $\frac{d\theta}{dz}$ we get:

$$\frac{d\theta}{dz} = \frac{(287 - 288)}{(100 - 40)} = -0.0167$$

so the atmosphere is unstable. Had the atmosphere been neutral, the potential temperature at 100 m would have been the same as at 40 m, that is, 288 K.

4.26 The expression of the wind profile in a stable atmosphere is given by:

$$u(z) = \frac{u_*}{\kappa} \left[\ln \left(\frac{z}{z_0} \right) + 5 \frac{z}{L} \right]$$

with the values given ($u_* = 0.4$, $z_0 = 0.15$, $L = 250$), the profile looks like:

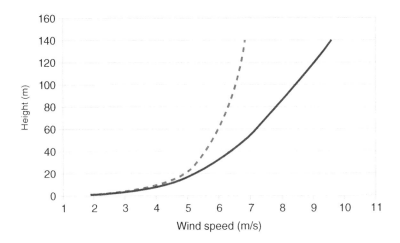

The dashed line is the profile for the neutral case (with $u_* = 0.5$, $z_0 = 0.15$).

4.27   The expression of the wind profile in an unstable atmosphere is given by:

$$u(z) = \frac{u_*}{\kappa}\left[\ln\left(\frac{z}{z_0}\right) - \left(2\ln\left(\frac{1+x}{2}\right) + \ln\left(\frac{1+x^2}{2}\right) - 2\tan^{-1}(x) + \frac{\pi}{2}\right)\right]$$

where $x = \left(1 - 16\frac{z}{L}\right)^{1/4}$

With the values given ($u_* = 0.4$, $z_0 = 0.15$, $L = -250$), the profile looks like:

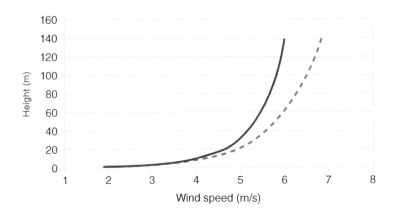

The dashed line is the profile for the neutral case (with $u_* = 0.5$, $z_0 = 0.15$).

4.28 Looking at Figure 4.12, it can be seen that out of the incoming 340.4 W/m², 99.9 W/m² is reflected back, which means that the albedo is:

$$\frac{99.9}{340.4} = 0.29$$

so, the Earth's albedo is 0.29, which is very similar to what others find.

4.29 Today the temperature outside is around 4°C where I am, that means had there been no greenhouse effect it would have been $(4 - 30) = -26°C$!

4.30 Iterating the geostrophic drag law with the values given ($G = 12$ m/s, $z_0 = 0.75$ m, $z = 100$ m, at 60°N):

$$G = \frac{u_*}{\kappa} \sqrt{\left[\ln\left(\frac{u_*}{fz_0}\right) - A\right]^2 + B^2}$$

it is possible to find that to get $G = 12$, u_* must be 0.581, and inserting this in the log profile we get:

$$u(100) = \frac{0.581}{0.4} \ln\left(\frac{100}{0.75}\right) = 7.1 \text{ m/s}$$

Chapter 5   Local Flow

5.1 Referring back to Table 2.2 in Chapter 2 we find that the microscale is characterised by time scales on the order of minutes to hours and length scales on the 100 m to around a kilometre scale.

5.2 Using Equation 5.1 and inserting A for the first area and $0.9A$ (i.e. a 10% reduction) for the second, and calling the unknown change S (i.e. u is changed to a wind speed of S times u), we get:

$$Au = 0.9ASu \Rightarrow S = \frac{1}{0.9} = 1.11$$

which means that the wind speed has increased by 11% due to the narrowing in of the tube by 10%.

5.3 When the flow passes over a valley, there is more 'space' for the streamlines, meaning that the distance between them increases, increasing the area, meaning again – according to Bernoulli's equation – that the speed is reduced.

5.4 Please see text following Exercise 5.4.

5.5 Doing this via a Google search (search phrase: 'detached flow wind') I found a lot of good ones, the list will of course change all the time, so I will not mention any here, so you don't have to click in vain.

5.6 Please see text following Exercise 5.6.

5.7 Please see Equation 4.9. As you can see, the roughness is used to 'normalise' the height. Higher roughness will give lower winds, given the same u_*.

5.8 The wind, u, is 7.5 m/s at 30 m $= z$. To go from one height to another, we need to use the log profile, and to isolate u_*, viz

$$u_* = \frac{u(z)\kappa}{\ln(z/z_0)}$$

inserting first $z_0 = 3$ cm and then $z_0 = 20$ cm, we get $u_* = 0.43$ and 0.60, respectively. Using these two values to calculate the wind speed at 100 m, we get: $u(100) = 8.8$ and 9.3, for the two z_0's.

5.9 Going back to the chapter on profiles (Chapter 4), particularly the logarithmic wind profile, you can see that if you have mast measurements from a minimum of two heights, you are able to solve the equation for z_0 (as well as for u_*). Please see the answer to Exercise 4.16 for the mathematical details.

5.10 The estimated roughnesses of the three cases are:

Covered in snow: $z_0 = 10^{-3}$ m
Corn: $z_0 = 0.1$ m
Just harvested (completely bare soil): $z_0 = 5 \cdot 10^{-3}$ m

5.11 Please see text following Exercise 5.11.

5.12 Please see text following Exercise 5.12.

5.13 Please see text following Exercise 5.13.

5.14 Using Table 2.2, we find that the mesoscale is characterised by lengths of 10^5 m and time scales on the order of hours.

5.15 The rule of thumb for the rate of change of the height of the IBL was that it grew as 1:100 (see Section 4.7).

5.16 We assume that it is only the sea breeze that determines the flow on the island, that means that the wind rose at the two locations will look as follows:

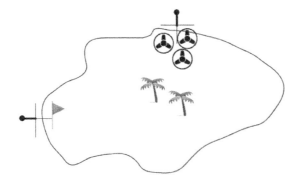

where the thick lines in the crosses mark the sectors of the rose, both are – as expected – unidirectional.

5.17 As in Exercise 5.2, we will use Bernoulli's equation (Equation 5.1) and inserting A for the first area and $1.15A$ (i.e. a 15% increase) for the second, and calling the unknown change S (i.e. u is changed to a wind speed of S times u), we get:

$$Au = 1.15ASu \Rightarrow S = \frac{1}{1.15} = 0.87$$

which means that the wind speed has decreased by 13% due to the widening of the tube by 15%.

5.18 Using Figure 5.9, and converting the two distances into multiples of the obstacle height, we get the reduction to be: reduction ($x = 400 = 40h$, $y = 15 = 1.5h$) equal to approximately 0.13, which means that the wind speed will be reduced by 13%.

5.19 Using Charnock's formula (Equation 5.6) with $u_* = 0.6$ m/s, we get:

$$z_0^w = a_c u_*^2/g = 0.016 \cdot 0.6^2/9.81 = 5.9 \cdot 10^{-4} \text{ m}$$

5.20 The roughness of a suburban area is around 0.5 m.

Chapter 6 Turbulence

6.1 Doing this via a Google search (search phrase: 'onset of turbulence') I found a lot of good ones, the list will of course change all the time, so I will not mention any here, so you don't have to click in vain.

6.2 Determine the unit of the Reynolds number.

$$\text{Re} = \frac{UL}{\nu} = \frac{m/s\,m}{m^2/s} = [\cdot]$$

i.e. no unit, so as expected an dimensionless number.

6.3 As typical values for atmospheric flow I have chosen 10 m/s as a characteristic wind speed and 500 m as a characteristic length scale (here the height of the boundary layer)

$$\text{Re} = \frac{UL}{\nu} = \frac{10 \cdot 500}{1.48 \cdot 10^{-5} \, m^2/s} = 34 \cdot 10^7$$

which is well above the threshold for turbulent flow (10^3). In fact, even for very low wind speeds and shallow boundary layers the atmospheric flow is still turbulent.

6.4

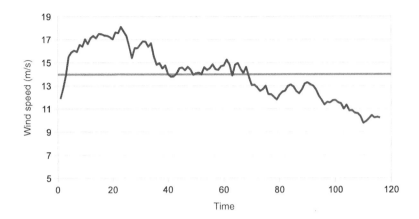

where the horizontal line marks the average.

6.5

$$\overline{u'} = \overline{u - \bar{u}} = \bar{u} - \bar{u} = 0$$

which is as expected, since the definition is that the mean is already removed.

6.6 Turbulence intensity is defined as:

$$\text{TI} = \frac{\sigma_u}{\bar{u}} = \frac{m/s}{m/s} = [\cdot]$$

so dimensionless. Often the number is multiplied by 100 to get a percentage.

6.7 The two main courses I can think of are wakes and forests. Some flow phenomena, as separated flow, might also create much higher turbulence in a given sector.

6.8 $\bar{u} = 10$ m/s; $\sigma_u = 2$ m/s:

$$\text{TI} = \frac{\sigma_u}{\bar{u}} = \frac{2}{10} = 0.2$$

$\bar{u} = 10$ m/s; $u_* = 0.3$ m/s:

using the relation in Box 20, we first find $\sigma_u = Au_* = 2 \cdot 0.3 = 0.6$ and then we can calculate TI:

$$\text{TI} = \frac{\sigma_u}{\bar{u}} = \frac{0.6}{10} = 0.06$$

$z = 60$ m; $z_0 = 0.1$ m:

Again using a relation from Box 20, we find

$$TI = \frac{1}{\ln\left(\frac{z}{z_0}\right)} = \frac{1}{\ln\left(\frac{60}{0.1}\right)} = 0.16$$

Note that the relations from the box we have used here should be used with great care.

6.9 Using the simple relation from Equation 6.6, we get:

$$u_{50}(u = 8) = 5 \cdot 8 = 40 \text{ m/s}$$

and

$$u_{50}(u = 10) = 5 \cdot 10 = 50 \text{ m/s}$$

Note, that this is indeed a very approximate relation.

6.10 $V_{ref} = 41$ m/s, TI = 0.15: **IIA**
$V_{ref} = 50$ m/s, TI = 0.13: **IB**
V_{ref} unknown, but $u = 9.5$, TI = 0.1: first we need to calculate V_{ref}, which can be found from Equation 6.6:
$V_{ref} = 5 \cdot 9.5 = 47.5$, so the class is **IC**

Chapter 7   Wakes

7.1 Please see text following Exercise 7.1.

7.2 Please see text following Exercise 7.2.

7.3 Please see text following Exercise 7.3.

7.4 Please see text following Exercise 7.4.

7.5 Please see text following Exercise 7.5.

7.6 Please see text following Exercise 7.6.

7.7 Please see text following Exercise 7.7.

7.8 Please see text following Exercise 7.8.

7.9 Please see text following Exercise 7.9.

7.10 This is a little bit of a difficult question to answer since, the effect of the wake will continue until infinity, getting smaller and smaller, but never zero. However, the value at 20 km is 9.99 m/s, so here we are very close to the undisturbed value.

7.11   Reading from Figure 7.12 we get:

cut-in: around 3 m/s
cut-out: 25 m/s
rated power: 1 MW

7.12   Here is a rough and ready calculation: I assume the following:

Number of turbines: 100
Rated power of turbines: 5 MW
Price per kWh: 0.1 EUR (bid price from the Horns Rev 3 wind farm)
Capacity factor: 0.40

This means that the annual production will be:

$$P = 100 \text{ turbines} \times 5000 \text{ kW/turbine} \times 8760 \text{ hours} \times 0.4 = 1{,}752{,}000{,}000 \text{ kWh}$$

which will cost:

$$C = 1{,}752{,}000{,}000 \times 0.1 \text{ EUR} = 175{,}200{,}000 \text{ EUR}$$

Five percent of this is then:

$$\text{Loss} = 0.05 \times 175{,}200{,}000 = 8{,}760{,}000 \text{ EUR}$$

which is almost 9 million EUR. Just to underline what this means: if I can shave off even half a percent of the wake loss, by for example, optimising the layout, I can make almost a million EUR.

7.13   Using for example, Google and possibly expanding the search term with 'off shore wind' should give you a good idea of where the industry is currently.

Chapter 8   Modelling

8.1   Referring back to Table 2.2, we find:

Global/planetary: length: 10^7 m; time: order of weeks
Synoptic: length: 10^6 m; time: order of days

8.2   Going to my local meteorological institute (DMI.dk) I found the following: an animation showing the output of the HIRLAM model (an NWP model) in 1 hour steps out to 2 days. Precipitation, wind, temperature, pressure, etc. was shown. A colour-coded scale was used to illustrate the magnitude of the various variables.

8.3

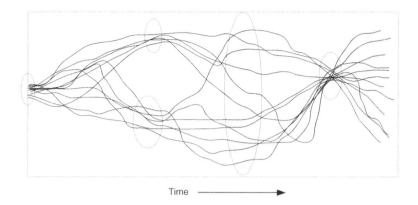

Time ──────────▶

Identification of the different areas of the ensemble prediction. There are points in time where the system seems to fall in one of two stages, others where it is very unclear, and at one time all the paths go through the same state. Time is along the x-axis, and a generalised state (think e.g. of wind speed) of the system along the y-axis.

8.4   Here is what I have found:

- German Weather Service (DWD): COSMO-EU, 7 km resolution, focus on 1–3 days ahead
- Meteorological Service of Canada: Regional Deterministic Prediction System (RDPS), 10 km (minimum) resolution, out to at least 48 hours
- Bureau of Meteorology, Australia: Australian Community Climate and Earth-System Simulator (ACCESS), 12 km (regional), out to at least 3 days.

and there are many more.

8.5   A good place to start is: http://reanalyses.org/atmosphere/erainterim-references

Chapter 9   Conclusion

9.1   Re-read the book as the first step; maybe just use the table of contents as a start – things people often forget to take into account include: forestry, especially the sublayer; recirculation bubbles; proximity to mesoscale effects, like coast lines and valleys, to mention but a few.

9.2   Over to you!

C

Sample Wind Speed and Direction Data

Sample of the data used extensively in Chapter 3, see the chapter for more details. The full data set can be downloaded from the book's website (QR code in margin). The first number is the time group (year, month, day, hour, minute), the second the wind speed and the third the direction. Note, that there are two sets of groups of numbers.

200101010005	7.42	10	200101010415	5.80	159
200101010015	6.69	10	200101010425	6.52	143
200101010025	6.79	10	200101010435	6.10	141
200101010035	7.55	10	200101010445	6.92	140
200101010045	7.58	10	200101010455	7.78	137
200101010055	8.04	10	200101010505	7.65	138
200101010105	8.24	10	200101010515	8.21	138
200101010115	7.62	10	200101010525	8.74	138
200101010125	7.35	10	200101010535	8.57	135
200101010135	7.32	10	200101010545	8.84	131
200101010145	7.25	11	200101010555	9.33	133
200101010155	7.55	118	200101010605	9.60	139
200101010205	7.58	125	200101010615	9.93	141
200101010215	7.52	125	200101010625	9.83	141
200101010225	7.19	126	200101010635	9.96	142
200101010235	7.09	126	200101010645	12.14	143
200101010245	6.39	130	200101010655	10.82	141
200101010255	7.05	137	200101010705	10.42	143
200101010305	7.19	138	200101010715	11.61	139
200101010315	8.14	139	200101010725	11.18	140
200101010325	7.42	145	200101010735	10.75	136
200101010335	6.79	158	200101010745	10.39	132
200101010345	6.19	163	200101010755	10.75	137
200101010355	6.92	163	200101010805	10.92	141
200101010405	5.90	162	200101010815	10.89	137

References

Abramovich GN. 1963. *The Theory of Turbulent Jets.* MIT Press, 671pp.

Ainslie JF. 1988. Calculating the flowfield in the wake of wind turbines. *Journal of Wind Engineering and Industrial Aerodynamics*, **27**, 213–224.

Amazon, 2015. aws.amazon.com (accessed 17 January 2015).

Ayotte KW, Davy RJ, Coppin PA. 2001. A simple temporal and spatial analysis of flow in complex terrain in the context of wind energy modelling. *Boundary-Layer Meteorology*, **98**(2), 275–295.

Banta RM, Pichugina YL, Kelley ND, Hardesty RM, Brewer WA. 2013. Wind energy meteorology: insight into wind properties in the turbine-rotor layer of the atmosphere from high-resolution doppler Lidar. *Bulletin of the American Meteorological Society*, **94**, 883–902.

Barthelmie RJ, Frandsen ST, Rathmann O, Hansen K, Politis E, Prospathopoulos J, Schepers JG, Rados K, Cabezón D, Schlez W, Neubert A, Heath M. 2011. Flow and wakes in large wind farms. Final Report for UpWind WP8 Risø-R-1765(EN).

Beck H, Trabucchi D, Bitter M, Kühn M. 2014. The Ainslie wake model-an update for multi megawatt turbines based on state-of-the-art wake scanning techniques. Proceedings from EWEA14, 10–13 March. Barcelona, Spain.

BIPM, 2015. www.bipm.org/en/bipm/mass/prototype.html (accessed 22 February 2015).

Bjerknes V. 1900. The dynamic principles of the circulatory movements in the atmosphere. *Monthly weather review*, **28**(10), 434–443.

Bleeg J, Digraskar D, Woodcock J, Corbett J-F. 2015. Modeling stable thermal stratification and its impact on wind flow over topography. *Wind Energy*, **18**(2), 369–383.

Bowen AJ, Mortensen NG. 1996. Exploring the limits of WAsP: the wind atlas analysis and application program. Proceedings of the 1996 European Union Wind Energy Conference and Exhibition, 20–24 May. Göteborg, Sweden, 584–587.

Burton T, Jenkins N, Sharpe D, Bossanyi E. 2011. *Wind Energy Handbook*, 2nd Edition, John Wiley & Sons, 780pp.

Businger JA. 1988. A note on the Businger–Dyer profiles. *Boundary-Layer Meteorology*, **42**(1–2), 145–151.

Charnock H. 1955. Wind stress over a water surface. *Quarterly Journal of the Royal Meteorological Society*, **81**, 639–640.

Clay Mathematics Institute. 2014. www.claymath.org/millennium-problems (accessed 30 November 2014).

Corbett J-F, Landberg L. 2012. Optimising the parameterisation of forests for WAsP wind speed calculations. European Wind Energy Association Conference (EWEA) 2012.

Dellwik E, Bingöl F, Mann J. 2013. Flow distortion at a dense forest edge. *Quarterly Journal of the Royal Meteorological Society*, **140**(679), 676–686.

Diebold M, Higgins C, Fang J, Bechmann A, Parlange MB. 2013. Flow over hills: A large-eddy simulation of the Bolund case. *Boundary-Layer Meteorology*, **148**(1), 177–194.

DMI. 2015. Hirlam model. www.dmi.dk/laer-om/temaer/meteorologi/hirlam/ (in Danish) (accessed 18 January 2015).

Dyer AJ. 1974. A review of flux-profile relationships. *Boundary Layer Meteorology*, **20**, 35–49.

Meteorology for Wind Energy: An Introduction, First Edition. Lars Landberg.

© 2016 John Wiley & Sons, Ltd. Published 2016 by John Wiley & Sons, Ltd.

ECMWF. www.ecmwf.int/en/forecasts/documentation-and-support (accessed 18 January 2015).

Ekman VW. 1905. On the influence of the Earth's rotation on ocean currents. *Arch. Math. Astron. Phys.*, **2**, 1–52.

Elliott WP. 1958. The growth of the atmospheric internal boundary layer. *Eos, Transactions American Geophysical Union*, **39**(6), 1048–1054.

Ferreira AD, Lopes AMG, Viegas DX, Sousa ACM. 1995. Experimental and numerical simulations of flow around two-dimensional hills. *Journal of Wind Engineering and Industrial Aerodynamics*, **54/55**, 173–181.

Garratt JR. 1992. *The Atmospheric Boundary Layer*. Cambridge University Press, ISBN 0-521-38052-9.

Ginzburg VL. 2001. *The Physics of a Lifetime: Reflections on the Problems and Personalities of 20th Century Physics*. Springer, 3–200.

Google, 2015. cloud.google.com (accessed 17 January 2015).

Gryning S-E, Batchvarova E, Brümmer B, Jørgensen H, Larsen S. 2007. On the extension of the wind profile over homogeneous terrain beyond the surface boundary layer. *Boundary-Layer Meteorology*, **124**(2), 251–268.

Hasager CB, Rasmussen L, Peña A, Jensen LE, Réthoré P-E. 2013. Wind farm wake: The Horns Rev photo case. *Energies*, **6**, 696–716.

Hasager CB. 2014. Offshore winds mapped from satellite remote sensing. *WIREs Energy and Environments*, **3**, 594–603.

Hasager CB, Nielsen M, Astrup P, Barthelmie R, Dellwik E, Jensen NO, Jørgensen BH, Pryor SC, Rathmann O, Furevik BR. 2005. Offshore wind resource estimation from satellite SAR wind field maps. *Wind Energy*, **8**(4), 403–419.

Hodgetts B, Harman K, Strachan A, Ebsworth G, Beaumont A. www.gl-garradhassan.com/assets/downloads/A_Statistical_Review_of_Recent_Wind_Speed_Trends_in_the_UK.pdf. (accessed 15 January 2015).

Hoskins BJ, Bretherton FP. 1972. Atmospheric frontogenesis models: mathematical formulation and solution. *Journal of the Atmospheric Sciences*, **29**, 11–13.

IEC. 2005. *Wind Turbines, Part 1: Design Requirements 61400-1*. IEC 61400-1:2005(E).

IEC. 2005. *Wind Turbines, Part 12-1: Power Performance Measurements of Electricity Producing Wind Turbines*. IEC 61400-12-1:2005.

Jackson PS, Hunt JCR. 1975. Turbulent wind flow over a low hill. *Quarterly Journal of the Royal Meteorological Society*, **101**, 929–955.

Jensen NO. 1983. A note on wind generator interaction. Risø M-2411.

Jensen NO, Petersen, EL, Troen I. 1984. Extrapolation of mean wind statistics with special regard to wind energy applications. WMO World Climate Programme Report No. WCP-86.

JMA, 2015. Global spectral model. www.jma.go.jp/jma/en/Activities/nwp.html (accessed 18 January 2015).

Kalnay E, Kanamitsu M, Kistler R, Collins W, Deaven D, Gandin L, Iredell M, Saha S, White G, Woollen J, Zhu Y, Leetmaa A, Reynolds R, Chelliah M, Ebisuzaki W, Higgins W, Janowiak J, Mo KC, Ropelewski C, Wang J, Jenne R, Joseph D. 1996. The NCEP/NCAR 40-year reanalysis project. *Bulletin of the American Meteorological Society*, **77**, 437–470.

Katic I, Højstrup J, Jensen NO. 1986. A simple model for cluster efficiency. Proceedings of the 1986 European Wind Energy Conference and Exhibition, October 7–9. Rome, Italy, EWEC'86.

Kristensen L. 1993. The cup anemometer and other exciting instruments. Risø-R-615(EN). Risø National Laboratory.

Landberg, L. 1993. Short-term prediction of local wind conditions. PhD thesis, University of Copenhagen.

Lettau H. 2011. A re-examination of the "Leipzig Wind Profile" considering some relations between wind and turbulence in the frictional layer. *Tellus A, North America*. **2**. www.tellusa.net/index.php/tellusa/article/view/8534 (accessed 24 July 2014).

Lorenz EN. 1963. Deterministic nonperiodic flow. *Journal of the Atmospheric Sciences*, **20**, 130–141.

Lorenz EN. 1972. Predictability: does the flap of a butterfly's wings in Brazil set off a tornado in Texas? Presented before the American Association for the Advancement of Science, December 29, 1972.

Mann J, Sathe A, Gottschall J, Courtney M. 2012. Lidar turbulence measurements for wind energy, in *Progress in Turbulence and Wind Energy IV*, edited by Oberlack M, Peinke J, Talamelli A, Castillo L, Hölling M. Springer, 263–270.

McIlveen R. 1986. *Basic Meteorology a Physical Outline*. Van Nostrand Reinhold (UK) Co Ltd, 457pp.

MEASNET, 2015. www.measnet.com (accessed 22 February 2015).

Mildner P. 1932. Uber Reibung in einer speziellen Luftmasse. *Beitr. Phys. fr. Atmosph*, **19**, 151–158.

Miyake M. 1965. Transformation of the atmospheric boundary layer over inhomogeneous surfaces. Science Report 5R-6. University of Washington.

Monin AS, Obukhov AM. 1954. Basic laws of turbulent mixing in the surface layer of the atmosphere. *Tr. Akad. Nauk SSSR Geofiz. Inst*, **24**, 163–187.

Mortensen NG, Heathfield DN, Myllerup L, Landberg L, Rathmann O. 2007. Getting started with WAsP 9. Technical Report Risø-I-2571(EN). Risø National Laboratory, 70pp.

Mortensen, NG, Heathfield DN, Rathmann O, Nielsen M. 2014. *Wind Atlas Analysis and Application Program: WAsP 11 Help Facility.* Department of Wind Energy, Technical University of Denmark.

NASA. 2014. Where does the Earth's atmosphere come to an end? http://image.gsfc.nasa.gov/poetry/ask/a10022.html (accessed 15 April 2014).

NASA. 2015. http://mars.nasa.gov/programmissions/missions/past/viking/ (accessed 11 January 2015).

NOAA. 2015a. www.ncdc.noaa.gov/data-access/model-data/model-datasets/global-forcast-system-gfs (accessed 18 January 2015).

NOAA. 2015b. www.weather.gov/ops2/ua/radiosonde/ (accessed 1 March 2015).

Nygaard NG. 2014. Wakes in very large wind farms and the effect of neighbouring wind farms. *Journal of Physics: Conference Series*, **524**. The Science of Making Torque from Wind 2014.

Obukhov AM. 1946. Turbulence in an atmosphere with a non- uniform temperature. *Tr. Inst. Teor. Geofiz. Akad. Nauk. SSSR*, **1**, 95–115.

Obukhov AM. 1971. Turbulence in an atmosphere with a non-uniform temperature (English translation). *Boundary-Layer Meteorology*, **2**, 7–29.

Ott S, Berg J, Nielsen M. 2011. Linearised CFD models for wakes. Risø-R-1772(EN).

Panofsky HA, Dutton JA. 1984. *Atmospheric Turbulence: Models and Methods for Engineering Applications.* John Wiley & Sons, 397pp.

Peña A, Floors R, Gryning S-E. 2014. The Høvsøre tall wind-profile experiment: a description of wind profile observations in the atmospheric boundary layer, *Boundary-Layer Meteorology*, **150**(1), 69–89.

Perera MDAES. 1981. Shelter behind two-dimensional solid and porous fences. *Journal of Wind Engineering and Industrial Aerodynamics*, **8**, 93–104.

Pope SB. 2000. *Turbulent Flows.* Cambridge University Press.

Raynor GS, Sethuraman S, Brown RM. 1979. Formation and characteristics of coastal internal boundary layers during onshore flows. *Boundary-Layer Meteorology*, **16**, 487–514.

Richardson LF. 2007. *Weather Prediction by Numerical Process*, 2nd edition, Cambridge University Press.

Rienecker MM, Suarez MJ, Gelaro R, Todling R, Bacmeister J, Liu E, Bosilovich MG, Schubert SD, Takacs L, Kim G-K, Bloom S, Chen J, Collins D, Conaty A, da Silva A, Gu W, Joiner J, Koster RD, Lucchesi R, Molod A, Owens T, Pawson S, Pegion P, Redder CR, Reichle R, Robertson FR, Ruddick AG, Sienkiewicz M, Woollen J. 2011. MERRA: NASA's modern-era retrospective analysis for research and applications. *Journal of Climate*, **24**, 3624–3648.

Salmon JR, Walmsley JL. 1986. User's guide to the MS3DJH/3R model. *Boundary Layer Research Division, Atmospheric Environment Service.*

Sathe A, Mann J, Vasiljevic N, Lea G. 2014. A six-beam method to measure turbulence statistics using ground-based wind lidars. *Atmospheric Measurement Techniques Discussion*, **7**, 10327–10359.

Sempreviva AM, Larsen SE, Mortensen NG, Troen I. 1990. Response of neutral boundary layers to changes of roughness. *Boundary-Layer Meteorology*, **50**(1–4), 205–225.

Skamarock WC, Klemp JB, Dudhia J, Gill DO, Barker DM, Duda MG, Huang XY, Wang W, Powers JG. 2008. A Description of the Advanced Research WRF Version 3. NCAR Technical Note.

Stoffelen A, Anderson DLT. 1993. ERS-1 scatterometer data and characteristics and wind retrieval skills. Proceedings of the First ERS-1 Symposium, ESA SP-359.

Taylor PA, Teunissen HW. 1987. The Askervein hill project: overview and background data. *Boundary-Layer Meteorology*, **39**(1–2), 15–39.

Tennekes H, Lumley JL. 1983. *A First Course in Turbulence.* The MIT Press.

Tijera M, Maqueda G, Yagüe C, Cano JL. 2012. Analysis of fractal dimension of the wind speed and its relationships with turbulent and stability parameters, Chapter 2 in *Fractal Analysis and Chaos in Geosciences*, edited by Sid-Ali Ouadfeul, Book under a creative commons licence.

Tillman JE, Landberg L, Larsen SE. 1994. The boundary layer of Mars: fluxes, stability, turbulent spectra, and growth of the mixed layer. *Journal of the Atmospheric Sciences*, **51**, 1709–1727.

Troen, I, Petersen EL. 1989. European wind atlas. Risø National Laboratory. Roskilde. ISBN 87-550-1482-8.

Troldborg N, Sørensen JN, Mikkelsen R. 2007. Actuator line simulation of wake of wind turbine operating in turbulent inflow. *Journal of Physics: Conference Series*, **75**. The Science of Making Torque from Wind.

Turner A. 2013. *The Indian Monsoon and Climate Change*. Walker Institute, University of Reading.

UK Met Office. 2015. Met Office Unified Model. www.metoffice.gov.uk/research/modelling-systems/unified-model/weather-forecasting (accessed 18 January 2015).

Uppala SM, Kållberg PW, Simmons AJ, Andrae U, Da Costa Bechtold V, Fiorino M, Gibson JK, Haseler J, Hernandez A, Kelly GA, Li X, Onogi K, Saarinen S, Sokka N, Allan RP, Andersson E, Arpe K, Balmaseda MA, Beljaars ACM, Van De Berg L, Bidlot J, Bormann N, Caires S, Chevallier F, Dethof A, Dragosavac M, Fisher M, Fuentes M, Hagemann S, Holm E, Hoskins BJ, Isaksen L, Janssen PAEM, Jenne R, Mcnally AP, Mahfouf J-F, Morcrette J-J, Rayner NA, Saunders RW, Simon P, Sterl A, Trenberth KE, Untch A, Vasiljevic D, Viterbo P, Woollen J. 2005. The ERA-40 re-analysis. *Quarterly Journal of the Royal Meteorological Society*, **131**(612), 2961–3012.

USA Today. 2014, http://usatoday30.usatoday.com/tech/science/columnist/vergano/2006-09-10-turbulence_x.htm (accessed 30 November 2014).

Volker P, Badger J, Hahmann AN, Ott S. 2012. Wind Farm parametrization in the mesoscale model WRF. European Wind Energy Association (EWEA 2012).

Weibull W. 1951. A statistical distribution function of wide applicability. *Journal of Applied Mechanics, Transactions ASME*, **18**(3), 293–297.

Wired. 2015. www.wired.com/2015/01/tech-time-warp-week-wwii-computers-rooms-full-humans/ (accessed 18 January 2015).

WHO. 2015. The Global Telecommunication System (GTS). www.wmo.int/pages/prog/www/TEM/GTS/index_en .html (accessed 18 January 2015).

WindFarmer 5.3. 2015. Theory Manual *DNV GL*.

Wood N. 1995. The onset of separation in neutral, turbulent flow over hills. *Boundary-Layer Meteorology*, **76**(1–2), 137–164.

World Meteorological Organization. 2008. WMO guide to meteorological instruments and methods of observation. Chapter 5: Measurement of surface wind. *World Meteorological Organization (WMO), Geneva*.

Index

Printed and bound by CPI Group (UK) Ltd, Croydon, CR0 4YY

15/06/2023

03227349-0001